教育部高等学校电子信息类专业教学指导委员会规划教材
高等学校电子信息类专业系列教材

微电子专业英语

陈铖颖 著

清华大学出版社
北京

内 容 简 介

集成电路及半导体核心技术作为现代信息社会的基础,是实现我国自主创新、自立自强的国之重器。集成电路设计作为集成电路产业重要的一环,关乎我国在集成电路领域的核心竞争力和地位。

为了满足国内高等院校微电子专业本科及研究生阶段专业英语的学习需求,本书以英文的形式介绍了 CMOS 集成电路设计的相关技术。全书共为三部分:第一部分为集成电路设计的基本知识,包括模拟集成电路设计以及数字集成电路设计的基本流程和 EDA 工具;第二部分介绍模拟集成电路设计的基本理论和电路,包括运算放大器、带隙基准源、低压差线性稳压源、模拟滤波器、比较器和模数转换器;第三部分简要阐述低功耗数字集成电路的设计方法。

本书可作为高等院校微电子学与固体电子学、集成电路设计相关专业的专业英语教材,也可供相关专业的工程技术人员使用。

本书封面贴有清华大学出版社防伪标签,无标签者不得销售。
版权所有,侵权必究。举报:010-62782989,beiqinquan@tup.tsinghua.edu.cn。

图书在版编目(CIP)数据

微电子专业英语/陈铖颖著. —北京:清华大学出版社,2021.6(2024.8重印)
高等学校电子信息类专业系列教材
ISBN 978-7-302-54790-7

Ⅰ.①微… Ⅱ.①陈… Ⅲ.①微电子技术－英语－高等学校－教材 Ⅳ.①TN4

中国版本图书馆 CIP 数据核字(2020)第 001672 号

责任编辑:	贾　斌
封面设计:	李召霞
责任校对:	梁　毅
责任印制:	沈　露

出版发行:清华大学出版社
  网　　址:https://www.tup.com.cn,https://www.wqxuetang.com
  地　　址:北京清华大学学研大厦 A 座　　　　　邮　　编:100084
  社 总 机:010-83470000　　　　　　　　　　　邮　　购:010-62786544
  投稿与读者服务:010-62776969,c-service@tup.tsinghua.edu.cn
  质量反馈:010-62772015,zhiliang@tup.tsinghua.edu.cn
  课件下载:https://www.tup.com.cn,010-83470236
印 装 者:三河市天利华印刷装订有限公司
经　　销:全国新华书店
开　　本:185mm×260mm　　印　张:12　　字　数:293 千字
版　　次:2021 年 8 月第 1 版　　　　　　　　　印　次:2024 年 8 月第 2 次印刷
印　　数:1501～2000
定　　价:49.00 元

产品编号:081302-01

前 言
PREFACE

在当今社会中,以信息技术为代表的高新技术突飞猛进,信息产业发展水平已成为衡量一个国家综合国力的重要标志。集成电路(Integrated Circuit,IC)作为当今信息时代的核心技术产品,其在国民经济、国防建设以及人们日常生活的重要性愈加突显。

针对国内高等院校微电子专业本科及研究生阶段专业英语的学习需求,本书以英文的形式介绍了 CMOS 集成电路设计的相关技术,并配以重点词、句的中英文对照,旨在逐步提高学生对微电子专业英文资料的阅读、理解和翻译能力,为今后更有效获取和交流专业信息打下坚实的基础。

本书主要分为三大部分内容,共七章内容。

第一部分:第 1 章 主要介绍 CMOS 模拟集成电路设计、数字集成电路设计的基本流程和 EDA 工具,并简要分析了 MOS 晶体管的电路模型和计算机仿真模型。

第二部分:第 2 章～第 6 章 介绍 CMOS 模拟集成电路设计的基本理论和电路,包括运算放大器、带隙基准源、低压差线性稳压源、模拟滤波器、比较器和模数转换器;对各类电路的结构、参数指标进行了详细的分析和阐述。

第三部分:第 7 章 简要分析了低功耗数字集成电路的设计方法,包括数字集成电路功耗的主要来源、门控时钟、门控电源以及多电源域等技术。

本书受"厦门理工学院教材建设基金资助项目"、福建省本科高校一般教育教学改革研究项目"FBJG 20180270"福建省新工科研究与改革实践项目、福建省教育科学"十三五"规划课题,厦门市教育科学"十三五"规划课题以及厦门市科技计划项目(青年创新基金项目)资助,由厦门理工学院微电子学院陈铖颖老师主持,陈黎明老师、张宏怡教授、北京理工大学信息与电子学院王兴华老师、天津大学微电子学院叶茂老师共同编纂完成。陈铖颖老师完成了第 1、2、3、6 章的编写,王兴华老师进行了第 4 章的编纂工作,第 5 章由叶茂老师完成,张宏怡教授和陈黎明老师共同完成了第 7 章的编写。同时张宏怡教授进行了全书的审校工作,正是有了大家的共同努力,才使本书得以顺利完成。本书涉及内容丰富,受时间和编者水平所限,书中难免存在不足和局限,恳请读者批评指正。

作 者
2021 年 5 月

目录
CONTENTS

Part I

Chapter 1　Basis of CMOS integrated circuit design ······ 3
  1.1　Analog versus Digital ······ 4
  1.2　CMOS analog integrated circuit design ······ 5
  1.3　CMOS analog integrated circuit EDA tool ······ 8
  1.4　CMOS digital integrated circuit design ······ 11
  1.5　CMOS digital integrated circuit EDA tool ······ 14
  1.6　CMOS device model ······ 16
    1.6.1　Large-signal model ······ 16
    1.6.2　Small-signal model ······ 17
    1.6.3　Noise model ······ 19
    1.6.4　Computer simulation model ······ 21
  1.7　Technical words and phrases ······ 24
    1.7.1　Terminology ······ 24
    1.7.2　Note to the text ······ 25

Part II

Chapter 2　Operational amplifier ······ 31
  2.1　Basis of OPA ······ 31
    2.1.1　Ideal OPA characteristics ······ 32
    2.1.2　Non-Ideal OPA characteristics ······ 32
  2.2　OPA structure ······ 33
    2.2.1　Telescopic cascode ······ 35
    2.2.2　Folded cascode ······ 35
    2.2.3　Gain-boosted ······ 36
    2.2.4　Two-stage ······ 38
  2.3　Feedback ······ 39
    2.3.1　Gain desensitization ······ 40
    2.3.2　Nonlinearity reduction ······ 40
    2.3.3　Bandwidth modification ······ 41
  2.4　Parameters ······ 41
  2.5　Industrial requirement of OPA design ······ 47
  2.6　Technical words and phrases ······ 48

    2.6.1 Terminology ································ 48
    2.6.2 Note to the text ··························· 49

Chapter 3 Bandgap and LDO ······························ 51
 3.1 Bandgap ··· 51
   3.1.1 Basic of bandgap ·························· 51
   3.1.2 Bandgap design ···························· 59
 3.2 Low-dropout linear regulator ······················ 62
   3.2.1 Basis of LDO ······························· 62
   3.2.2 Operational principle ····················· 63
   3.2.3 Parameter ···································· 64
   3.2.4 Stability analysis ·························· 69
 3.3 Technical words and phrases ······················· 72
   3.3.1 Terminology ································ 72
   3.3.2 Note to the text ··························· 73

Chapter 4 Analog filter ······································ 75
 4.1 The classification of analog filters ·············· 75
   4.1.1 According to the components ········· 75
   4.1.2 According to the functions ············· 76
   4.1.3 According to the approximation functions ············· 77
 4.2 Amplitude-frequency characteristics ············· 81
 4.3 Transfer function ···································· 82
   4.3.1 Low-pass transfer function ············· 83
   4.3.2 High-pass transfer function ············ 83
   4.3.3 Band-pass transfer function ············ 84
   4.3.4 Band-stop transfer function ············ 85
   4.3.5 All-pass transfer function ··············· 85
 4.4 Implementation of analog filter circuit ········· 87
   4.4.1 Active RC integrator ····················· 87
   4.4.2 MOS-C integrator ························ 87
   4.4.3 Gm-C integrator ·························· 88
   4.4.4 Active resistor ···························· 90
   4.4.5 Active inductor ··························· 91
 4.5 The realization method of analog filter ········ 92
   4.5.1 Cascade design method ················· 92
   4.5.2 LC ladder synthesis ······················ 93
 4.6 Complex filter ·· 94
 4.7 Parameter ·· 100
 4.8 Technical words and phrases ····················· 101
   4.8.1 Terminology ······························ 101
   4.8.2 Note to the text ························· 102

Chapter 5 Comparator ······································ 103
 5.1 Basis of comparator ······························· 103
 5.2 Parameter ·· 104

5.3 Characteristic analysis ·· 106
5.4 Comparator structure ·· 107
5.5 Basis of schmitt trigger ·· 111
5.6 Technical words and phrases ·· 116
 5.6.1 Terminology ·· 116
 5.6.2 Note to the text ·· 116

Chapter 6 Analog-to-Digital Converter ·· 118
6.1 Summary ·· 118
6.2 Performance parameters ·· 119
 6.2.1 Static parameter ·· 120
 6.2.2 Dynamic parameter ·· 122
6.3 Flash ADC ·· 124
 6.3.1 Flash-interpolation structure ······································ 125
 6.3.2 Folding-interpolation structure ··································· 126
 6.3.3 The main application areas of Flash ADC ··················· 127
6.4 Two-step ADC ·· 129
6.5 Folding ADC ·· 129
6.6 Interpolating ADC ·· 130
6.7 Successive Approximation Analog-to-digital Converter (SAR ADC) ··· 131
6.8 Pipelined ADC ·· 134
 6.8.1 Sample and hold ·· 137
 6.8.2 Sub-ADC ·· 139
 6.8.3 Comparator ··· 140
 6.8.4 MDAC ··· 142
6.9 Sigma-delta ADC ·· 144
 6.9.1 Oversampling ·· 146
 6.9.2 Noise shaping ·· 148
 6.9.3 Decimation filter ·· 151
 6.9.4 Parameter ··· 152
 6.9.5 The basic structure of Sigma-delta modulator ·············· 153
6.10 Technical words and phrases ··· 162
 6.10.1 Terminology ··· 162
 6.10.2 Note to the text ··· 163

Part Ⅲ

Chapter 7 Low Power Design for Digital CMOS Circuits ················ 167
7.1 Sources of Power Consumption ··· 167
7.2 Low Power Design Methodologies ··· 170
7.3 Physical Level of Low-Power Design ····································· 177
7.4 Technical words and phrases ·· 181
 7.4.1 Terminology ··· 181
 7.4.2 Note to the text ·· 182

参考文献 ·· 184

Part I

Part 1

Chapter 1

Basis of CMOS integrated circuit design

In today's society, Integrated circuit(IC), as an important pillar of the information industry, plays an increasingly important role in the national economic and social development, and has become an important symbol of a country's comprehensive national strength. Integrated circuit chip and software are the foundation of our information industry. The silicon material is created by design and process, and integrates information collection, processing, operation, transmission and storage functions into a single chips, which has become the foundation of information technology in human society.

The development of integrated circuit technology has gone through several stages. The SSI(Small Scale Integrated circuits) developed in the late 1950s has only 100 components. In 1960s, the MSI(Medium Scale Integrated circuits) was developed, with an integration of about 1000 components. In 1970s, the LSI(Large Scale Integrated circuit) was developed, and the integration was up to thousands of components. Quickly at the end of the 1970s, the VLSI(Very Large Scale Integrated circuits) came out and the integration degree reached about 10^5 components. In 1980s, ULSI(Ultra Large Scale Integrated circuits) entered the market on a large scale whose integration level increased by an order of magnitude over VLSI, reaching 10^6 components.

Similarly, the integrated circuit technology has also experienced from simple to complex development history, began appearing on the P-channel silicon-gate metal-oxide-semiconductor(MOS) technology, P-channel aluminum-gate metal-oxide-semiconductor technology, N-channel silicon-gate metal-oxide-semiconductor technology, high performance short-channel metal-oxide-semiconductor(HMOS) technology. They have their own advantages and disadvantages, applied in different periods and different areas.

With the increasing IC integration, the ordinary MOS technology can no longer meet the needs of large-scale and ultra large scale integrated system manufacturing, so the complementary metal-oxide-semiconductor (CMOS) process arises at the historic moment. Although the CMOS process is more complex than the NMOS process and the performance of the early CMOS devices is poor, the CMOS device has an absolute advantage in power consumption and integration. The manufacture of digital LSI and VLSI integrated circuits with CMOS can solve the most urgent problem of power consumption. Therefore, it has been widely applied and has been developing rapidly. Especially since 1980s, CMOS has become the leading manufacturing technology of VLSI such as Central Processing Unit(CPU), Random-Access Memory(RAM), Digital Signal Processor(DSP) and Microcontroller Unit(MCU). Its application scope has penetrated into every aspect of people's life.

1.1 Analog versus Digital

Integrated circuit is divided into two categories: analog and digital. Analog circuits are dealing with signals free to vary from zero to full power supply voltage. It means analog integrated circuit can process time-varying signal that can take on any value across a continuous range of voltage, current, or other metric. Therefore analog integrated circuits are usually used to deal with analog signals in nature, such as pressure, sound, wind speed, and so on. In all electronic systems, we need to first use analog circuits to convert voltage or current signals to digital quantities for further processing.

Different from analog integrated circuits, digital integrated circuits almost exclusively employ "all or nothing" signals: voltages restricted to ground and full supply voltage, with no valid state in between those extreme limits. A digital signal is modeled as taking on, at any time, only of two discrete values, which we call "0" and "1"("0" stands for ground voltage and "1" for supply voltage).

Digital signal processing algorithms were becoming more and more powerful with advances in integrated circuit technology. Many functions that had traditionally been realized in analog form were now easily performed in the digital domain, suggesting that, with enough capability in IC fabrication, all processing of signals would eventually occur digitally. However, in many of today's complex, high-performance systems, analog circuits have proved fundamentally necessary. Signals are analog occurring in nature at a macroscopic level. For example, at the receive end of optical receivers, the light is converted to a small electrical current by a photodiode. Ultrasound systems use an acoustic sensor to generate a voltage proportional to the amplitude of the ultrasound waveform. Amplification, filtering, analog-to-digital conversion, digital-to-analog conversion are fundamental functions in these applications.

But we still have to realize that in the past 20 years, digital integrated circuits have been applied in more areas and spread to many aspects of life. There are in fact many reasons to favor digital integrated circuits over analog ones:

(1) Reproducibility of results. Given the same set of inputs, a properly designed digital circuit always produces exactly the same results. The outputs of an analog circuit vary with temperature, power supply voltage, component aging and other factors.

(2) Ease of design. Digital design is logic design. Its process is similar to language expression. No special math or circuit skill are needed, and the behavior of small logic

circuits can be visualized mentally without any special insights about the operation of capacitors, transistors, or other devices that require calculus to model.

(3) Programmability. Most of us have already been quite familiar with digital computer and the ease with which we can design, write and debug programs. So the digital circuit design is similar, which is carried out by hardware description language (HDL), such as Verilog or VHDL. These languages allow both structure and function of a digital circuit to be specified or modeled. Besides a complier, a typical HDL also comes with simulation and synthesis programs. These software tools are used to test the hardware model's behavior before any real hardware is built, and then synthesize the model into a circuit in a particular component process.

(4) Speed. Today's transistor are very fast. With them a complete, complex system can operate in more than 10 gigahertz frequency. This means that such a system can produce 10 billion or more results per second.

(5) Economy. Digital circuits can provide a lot of functionality in a small chip. Circuit that are used repetitively can be integrated into a single chip and mass-produced at very low cost, making cheap products such as calculators, digital watches and mobile phones.

(6) Steadily advancing technology. Digital circuits always advance the progress of the technology. At present, the mainstream digital circuit technology has reached 28nm. And there will be faster, cheaper technology in a few years. So when designing a digital circuit, designers can accommodate these expected advances during the initial stage, to forestall system obsolescence and to add value for customers.

1.2　CMOS analog integrated circuit design

Analog circuit design technique, as the most classic and traditional art form in engineering technology, is still an irreplaceable design method in many complex high performance systems. The biggest difference between CMOS analog integrated circuit design and traditional discrete-component analog circuit design is that all active and passive devices are made on the same substrate, and the size is extremely small, which can not be verified by circuit board. Therefore, the designer must use the method of computer simulation to verify the circuit performance. Analog integrated circuit design consists of several stages. Fig.1.1 represents its general flow.

(1) Specification definition
(2) Circuit design
(3) Circuit pre-simulation
(4) Layout design
(5) Physical verification
(6) Parasitic extraction and post-simulation

(7) Exporting design data and tapeout

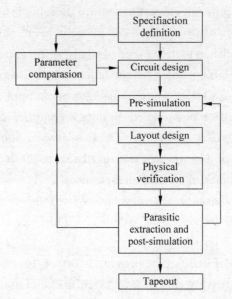

Fig. 1.1　CMOS analog integrated circuit design flow

A design flow begins with the system specification definition, and the designer should make clear the specific requirements and performance parameters of the design at this stage. After completing the circuit design, the next step is to evaluate circuit performance by computer simulation. At this time, the circuit may be further improved according to the simulation results, and the simulation can be repeated many times. Once the simulation results of the circuit performance can meet the design requirements, another main design work is required for the circuit layout. After layout completion and physical verification, the Parasitic of layout and wiring need to be taken into account again for computer simulation. If the simulation results are also satisfied with the design requirements, the design data can be output for tapeout.

Unlike using discrete devices to design analog circuits, the analog integrated circuit design can not be carried out in the way of building a circuit board. With the development of Electronic Design Automatic(EDA) technology, the above design steps are carried out by computer aided. By computer simulation, the signal can be monitored at any point in the wire the feedback loop can be opened; the circuit can be easily modified. But there are some restrictions on computer simulation. For example, the model is not perfect, and the program solution can not get the result due to the non convergence. The phases of the design process will be described in detail below.

1) Specification definition

In this stage the engineer should consider the whole system and its subsystem as a "black box" with only the relationship between input and output, not only to define each function, but also the timing, power consumption, area and Signal-to-Noise Ratio(SNR) and other performance parameter requirements.

2) Circuit design

According to the design requirements, the designer first chooses the appropriate process library(Process Design Kit,PDK) that is provided by each foundry,and then the rational frame system. Due to the complexity and diversity of CMOS analog integrated circuits,there are no EDA vendors that can fully solve the design automation of CMOS analog integrated circuits,so all the analog circuits are still designed by hand.

3) Circuit pre-simulation

The design engineer must confirm that the design is correct,for this purpose,based on the transistor model,the performance of the circuit is evaluated and analyzed with the aid of the EDA tool. At this stage,the transistor parameters should be modified according to the results of the circuit simulation. According to the change of the parameters in PDK,designer determines the interval and limitation of the circuit operation and verifies the influence of the environmental factors on circuit performance. Finally, simulation results will guide layout design as well.

4) Layout design

The design and Simulation determine the composition of the circuit and the related parameters,but it can not be directly sent to foundry for production. The design engineer needs to provide the physical geometric description of the integrated circuit,that is,the general "layout". This link is to convert the designed circuit into a graphic description format. CMOS analog integrated circuits are usually handmade layout designs with full custom methods. In design process,the effects of design rules,matching,noise,crosstalk, parasitism on the performance and manufacturability of the circuit are considered. Although many advanced full custom aided design methods have emerged, it is still impossible to guarantee the comprehensiveness of manual design for layout and various effects.

5) Physical verification

Does the layout design meet the manufacturing reliability requirements? Are there any new errors from the conversion of circuit to layout? The physical verification phase will solve the above two kinds of validation problems through the design rule check (DRC) and the comparison between the layout netlist and the circuit schematic(Layout Versus Schematic,LVS). The geometric rule check is used to ensure the feasibility of the layout in the process. It takes a given design rule as a standard to check the minimum line width,the minimum graphic spacing,the hole size,the minimum overlap area of gate and source,drain area, and so on. The comparison of the layout netlist and the circuit schematic is used to ensure the matching of the layout and the circuit design. The LVS tool extracts the circuit netlist from the layout,including the transistors and electrical connection size,and then compares it with the circuit netlist obtained by the schematic to check whether the two are consistent.

6) Parasitic extraction and post-simulation

The simulation before layout is an ideal simulation, which does not contain the parasitic parameters from layout, is called "pre-simulation". The simulation including parasitic information in layout is called "post-simulation". CMOS analog IC is more sensitive to parasitic parameters than digital IC. So even the results of pre-simulation meet the design requirements, which does not mean that the post-simulation can also be satisfied. In deep-submicron and nanoscale stage, the parasitic effect is more obvious, so the post-simulation analysis will be particularly important. Like pre-simulation, the transistor parameters, even some circuit structures, should be modified when the results are not satisfied in post-simulation. For high performance design, this process needs to be repeated many times, until the post-simulation meets the system design requirements.

7) Exporting designdata and tapeout

Passing post-simulation, the last step of design is to export the layout data (GDSII file), and submit it to foundry for production.

1.3 CMOS analog integrated circuit EDA tool

From the previous section, we can know that in CMOS analog integrated circuit design, circuit design and pre-simulation, layout implementation and layout physical verification, Parasitic extraction and post-simulation are the three most important steps for engineers. This section introduces several kinds of EDA design tools that are currently widely used in accordance with the design process.

1. Circuit design and simulation tools

The traditional tools of circuit design and simulation are mainly Cadence Spectre, SYNOPSYS HSPICE and Mentor Eldo. In addition, based on the above tools, in order to meet the needs of large-scale and rapid simulation, the three companies developed corresponding fast circuit simulation tools respectively, namely, Cadence Spectre Ultrasim, SYNOPSYS HSIM and Mentor Premier.

1) Spectre

Spectre is developed by Cadence for analog, mixed-signal and RF integrated circuit design and simulation. With powerful simulation function, it includes DC Analysis, Transient Analysis, AC small signal simulation (AC Analysis), pole zero analysis (PZ Analysis), noise analysis, Periodic Steady-State (PSS) Analysis and Mento Carlo Analysis etc. It can analyze and optimize the design yield, and greatly improve the design efficiency of the complex integrated circuit. In particular, its schematic input with graphical interface makes it the most commonly used CMOS analog integrated circuit design tool. Cadence also cooperate with major semiconductor manufacturers worldwide to establish a process library file—PDK, the designer can easily use different technology nodes for design and simulation. In addition, Spectre also provides co-simulation functions with other tools, such as HSPICE, ADS, and MATLAB. Coupled with the built-in rich element model base, it greatly increases convenience, rapidity and accuracy of the analog integrated circuit design.

2) HSPICE

HSPICE is an analog and mixed-signal integrated circuit design tool developed by original Meta-Software (now belonging to SYNOPSYS company). Unlike Cadence Spectre graphical input interface, HSPICE is simulated by reading circuit netlist and control statement. It is the most widely accepted simulation tool with the highest accuracy. Similar to Spectre, HSPICE also contains functions of DC simulation, transient simulation, AC small signal simulation, pole zero analysis, noise analysis, FFT analysis, worst case analysis and Monte Carlo analysis. In early version, when circuit was large or complex, the simulation matrix did not converge with HSPICE. After being acquired by SYNOPSYS, the problem was gradually improved through multiple editions. After the 2007sp1 version, HSPICE has made a great leap, and the problem of simulation convergence is also basically solved.

3) Eldo

Eldo is an EDA design tool developed by Mentor. Eldo can be simulated using the same command line as HSPICE, and can also be integrated into the circuit editing tool environment, such as DA_IC of Mentor, or Cadence Spectre.

In Eldo the global check is carried out by the Kirchhoff current constraint, and the convergence is strictly controlled and the same precision as HSPICE is ensured. Compared with early HSPICE version, the simulation speed of Eldo is faster. In terms of simulation convergence, Eldo adopts the concept of segmentation, so that the convergence of circuit is greatly improved.

Eldo can be easily embedded into other analog integrated circuit design environment, and it can be extended to hybrid-simulation platform ADMS for digital and analog co-simulation. The Eldo output files can be viewed and calculated by other kinds of wave observation tools. The Xelga and EZWave provided by Eldo are two powerful tools for wave observation and processing.

Due to the emergence of large-scale mixed-signal circuit and SoC, traditional analog integrated circuit simulation tools meet the bottleneck: mainly reflected in the slow speed, limited capacity (maximum support 50,000~100,000 devices). Major companies have developed a new generation of fast simulation tools. Usually, in order to improve the speed of simulation these tools mainly include: model linearization, model tabulation, multi-rate simulation, matrix segmentation, event driven technology and so on. Cadence's Spectre UltraSim, SYNOPSYS's HSIM, and Mentor's Premier are the best of them.

2. Layout tools

The layout tool is currently a unique situation of Cadence Virtuoso Layout Editor.

Only the Laker of SYNOPSYS has certain competitiveness.

1) Virtuoso Layout Editor

As an important product of Cadence's physical layout tools, Virtuoso Layout Editor is the most widely used layout tool at present. It can identify different process layer information and support customized ASIC, mixed-signal and analog design. It also adopts spatial-wiring technology and cooperates with other software components to completes the layout design quickly and accurately.

Virtuoso Layout Editor is mainly characterized by the following aspects:

(1) speed up customized layout at device, unit and module level.

(2) implementation of physical layout driven by constraint and circuit schematic.

(3) when designers submit schematic or need to evaluate and modify standard units, the function of fast standard cell can increase the layout performance by 10 times.

(4) provides constraint driven execution of advanced technology node and design rules.

2) Laker

Laker is a new generation of layout editing tool developed by Taiwan SprintSoft company. It was acquired by SYNOPSYS company in 2012, and now it has become an EDA layout tool of SYNOPSYS. Compared with the traditional Virtuoso layout tools, the biggest highlight of Laker is the creative introduction of Schematic Driven Layout technology, that is, the transformation function of circuit schematic similar to printed circuit board(PCB) EDA tools. The designer can directly import through schematic to form the layout, and get the pre-drawing line between the devices, which greatly reduces the errors caused by artificial layout and improves the editing efficiency. In addition, Laker has the following features:

A. The schematic window and layout window are displayed at the same time to facilitate the designer to see the devices and connections in real time.

B. Automatic layout, which quickly puts the device in the circuit map to a more suitable position.

C. Real time electrical rules checking, highlighting the operating layout elements, avoiding common faults, short and open.

3. Physical verification and Parasitic extraction tool

The physical validation consists of three parts, namely, DRC(Design Rule Check), LVS(Layout VS Schematic), and PEX(Parasitic Extraction). DRC mainly checks the layout rules, and also checks some DFM(Design For Manufacture), such as metal density, antenna effect, to ensure the requirement of process. LVS compares layout and schematic to ensure the consistency of design. PEX is used to extract parasitic parameters. Because schematic design doesn't considered metal wire and the device parasitic information, which(especially for deep-submicron design) will seriously affect timing and function. So after parasitic extraction, taking these factors into account post-simulation must be executed to ensure the success of design.

Similar to design and simulation tools, physical verification and Parasitic extraction tool also appeared in Cadence, SYNOPSYS and Mentor three companies. Assura, Hercules, and Calibre are physical verification and Parasitic extraction tools for Cadence, SYNOPSYS and Mentor respectively. In early process, Cadence has another command-line physical verification tool Dracula, which is now basically out of date. Moreover, compared to Assura and Calibre, the application of Hercules in CMOS analog layout verification is limited, and it is not introduced here.

1) Assura

Assura is regarded as an upgraded version of Diva in Spectre. It sets up a number of rules to support the physical verification, interactive and batch mode of large scale circuits. But before verification is done, the designer needs to manually export the netlist of schematic and layout. New version can open circuit and layout in the same interface, which greatly facilitates the designer to locate and modify DRC and LVS errors in layout. Parasitic extraction supports netlist format in Spectre, HSPICE, and Eldo environments. Designers can choose either of them for simulation

2) Calibre

Calibre is the most widely used EDA tool for physical verification in deep-submicron and nanoscale design. It can be easily embedded into the layout tools such as Virtuoso and Laker. Calibre uses graphical interface, and provides fast and accurate DRC, electrical rule check(ERC) and LVS function.

The hierarchical architecture of Calibre effectively simplifies difficulty of complex ASIC/SoC physical verification. Designers do not need special settings for different design. At the same time, they can browse and locate errors quickly and accurately according to physical verification results. The parallel processing capability of Calibre supports multi-core CPU operations, which can significantly shorten verification time for large scale design.

1.4 CMOS digital integrated circuit design

With the rapid development of semiconductor manufacturing and integrated circuit design technology, traditional principle schematic design method based on experience no longer meets the needs of times. With portability and independent characteristics of semiconductor process, the hardware description language (Hardware Description Language, HDL) is proposed. The design process of CMOS digital integrated circuit is mainly divided into two parts: front-end design and back-end design as Fig. 1. 2 shows. The front-end design mainly includes function-structure analysis and design, RTL coding and simulation, synthesis, gate-level simulation and pre-layout static timing analysis. Back-end design consists of place and route, post-layout static timing analysis and post-simulation.

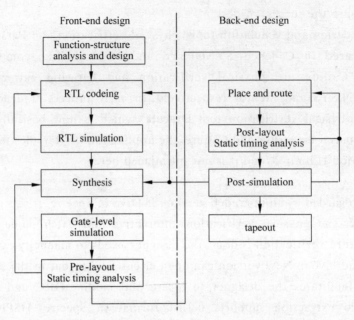

Fig. 1. 2　CMOS digital integrated circuit design flow

　　The main work of front-end design is to convert the circuit function into a hardware description language, and then synthesize code as a gate-level circuit. The back-end design mainly completes the layout. And post-simulation is aim to test the influence of the delay produced by the circuit after place and route, and verify whether the circuit meets design requirements of function and timing.

　　1) Function-structure analysis and design

　　Similar to analog integrated circuit design flow, function-structure analysis and design plays a very important role in the whole digital integrated circuit design. The design of the stage is based on product function definition, and preliminarily evaluate the process, power consumption, area, performance and cost, so as to make the corresponding design plan.

　　2) RTL coding

　　RTL coding uses the hardware description language(Verilog or VHDL) to describe the corresponding circuit design or behavioral function. Using a HDL to describe function and to write a test module, a good code style should have the following features: Sufficient annotation instructions and meaningful naming; it is not necessary to use non-blocking in combinatorial logic to improve readability and maintainability by adopting parameter definitions. pay attention to the width of the vector, and the width of the vector should also indicate the width of the value. the principle that the code can be synthesized.

　　3) RTL simulation

　　RTL simulation is used to check whether the design function meets the requirements.

After RTL code is completed, the correctness of the design must be tested. The usual method is to motivate the design module, and to test the correctness of the function by observing its output waveform. The test module is generally called test bench. In simulation, different test benches can be written to verify the design in all aspects.

4) Synthesis

Synthesis is through the logical synthesis tool to translate, optimize and map the design described by HDL, and generate the netlist related to semiconductor process. A netlist is a document that records the connection between logical gates and delay information. Synthesis is the bridge between RTL code and physical realization. Synthesis results determines the design of the circuit. The gate-level netlist after synthesis and given constraints will be sent to the back-end tool for place and route.

5) Gate-level simulation

Before layout we needs to simulate to check the design functions whether to meet the requirements. At this stage, the synthesized netlist is added to the simulation file and the technology library is needed to simulate. Gate-level simulation is more realistic than RTL-level simulation, because gate-level simulation takes into account the connection and delay information between logic gates.

6) Pre-layout static timing analysis

Static Timing Analysis(STA) is an important step to ensure that the circuit meets the scheduled timing requirements by analyzing specific timing models and whether they violate the designer's given timing constraints. But STA at this stage does not contain the connection delay information, it is only a preliminary timing verification.

7) Place and route

The designer should first carry out a reasonable layout plan to effectively use the resources to complete place and route. The specific work of layout planning is to calculate the size of each circuit module and to arrange their relative positions. With the improvement of technology, the role of connection is becoming more and more important, and the layout planning has more and more influence on the final design results. Then place and route will complete the placement and interconnection of different circuit units. Sometimes the division of planning, place and route is not very independent. Some EDA tools in practice may combine these steps together.

8) Post-layout static timing analysis

Different from Pre-layout static timing analysis, the static timing analysis at this stage needs to add wiring information. At this time, the delay between circuits will increase greatly. Accordingly, timing constraints set by the designer will also be severely challenged. So only through this verification can the designer ensure that the chip meets the default timing requirements.

9) Post-simulation

The gate-level simulation with delay information is the last gateway of design, mainly including functional and timing verification. Because functional verification is basically guaranteed in RTL-level simulation, this verification is mainly for timing simulation. Timing simulation needs to know some delay models adopted by simulation tools, and add the delayed information to simulation files. So that we can simulate some situations that can not appear at RTL-level, such as reset, state machine flip and so on. Full verification should include the best case of the shortest path hold time, and the longest path set-up time in the worst case.

1.5 CMOS digital integrated circuit EDA tool

From the last section, we can know that the design of digital IC mainly use EDA tools on RTL simulation, synthesis, static timing analysis and place & route. In this field, Cadence, SYNOPSYS and Mentor is also a situation of tripartite confrontation trend, so this section introduces the three companies' digital EDA design tools.

1. RTL simulation tool

At present, the mainstream RTL simulation tools include Mentor's ModelSim, SYNOPSYS's VCS, Cadence's NC-Verilog, Altera's Quartus II and Xilinx company's ISE.

1) ModelSim

In the field of RTL simulation, Mentor's ModelSim is the most widely used HDL simulation software in the industry. It provides a friendly simulation environment and is the only one of the industry's single kernel emulators that support VHDL and Verilog hybrid simulation. ModelSim adopts directly optimized compilation technology and single kernel simulation technology. Its compilation and simulation speed is fast. Compiled code is independent of platform. It is the preferred simulation software for digital integrated circuit designers.

ModelSim can perform simulation of behavior-level, RTL-level and gate-level separately or at the same time, and integrates many debugging functions such as performance analysis, waveform comparison, code coverage, virtual object, Memory window, source code window and so on. It also adds direct support to the SystemC so that it can be mixed with the HDL arbitrarily.

2) VCS(Verilog Compiled Simulator)

VCS is a compiled Verilog simulator for SYNOPSYS, which fully supports the Verilog HDL language of OVI standard. VCS has high simulation performance. Memory management capability can support tens of millions of ASIC design, and its simulation accuracy also meets the design requirements of deep-submicron ASIC.

VCS can be easily integrated into a test platform for Verilog, SystmVerilog, VHDL, and Openvera to generate bus communication and protocol violation check. At the same time, the monitor provides a comprehensive report to display the functional coverage of the bus communication protocol. The verification IP of the VCS library is also included in the DesignWare library, and can also be embedded as an independent tool suite.

3) NC-Verilog

NC-Verilog is an upgraded version of Cadence's original RTL simulation tool—Verilog-XL. Compared with the latter, NC-Verilog's simulation speed, huge design ability, and the storage capacity are greatly increased. When NC-Verilog is compiled, the Verilog code is first converted to a C program, and then the C program is compiled to the simulator. It is compatible with most of Verilog-2001 standards and has been updated by the Cadence. At present, in 64-bit systems, NC-Verilog can support more than 100 million gates of chip design.

2. Synthesis tool

For Synthesis tools, it is now a unique situation of SYNOPSYS company's DC (Design Compiler). In recent years, Mentor has also developed its own Synthesis tool, RealTime-Designer, but the market share is far less than DC.

DC(Design Compiler)

SYNOPSYS's DC is currently supported by more than 60 semiconductor manufacturers and more than 380 process libraries around the world, occupying nearly 91% of the market share. DC is the synthesis tool of industrial standards for more than 10 years, and is also the most core product of SYNOPSYS. According to design description and constraint conditions, an optimized gate-level circuit is automatically synthesized for a specific process library. It can accept multiple input formats, such as HDL, schematic and netlist, and produce multiple performance reports, which can shorten design time and improve design performance.

New version DC also extends topology technology to accelerate the adoption of advanced low power and test technology design convergence, helping designers improve production efficiency and chip performance. Topology technology can help designers correctly evaluate the chip power in synthesis process, and solve all the power problems in the early design. New DC has adopted a number of innovative synthesis technologies, such as adaptive retiming and clock gating. The performance increases by 8% compared with previous version, and the area reduces by 4%, power consumption reduces by 5%.

3. Static timing analysis tool

SYNOPSYS's PrimeTime is the only common tool for static timing analysis. PrimeTime is a standard gate-level static timing analysis tool. It can analyze the design of up to 500 million transistors at 28nm or even small technology nodes. In addition, PrimeTime also provides extended timing analysis, on-chip variable analysis, delay calculation and advanced modeling technology, and supports transistor models for most semiconductor manufactures. The new version of PrimeTime also includes PrimeTime SI, PrimeTime ADV and PrimeTime PX components, respectively, which provide analysis for signal integrity, on-chip variable and gate-level power and greatly accelerate design process.

4. Place & route tool

SYNOPSYS IC Compiler(ICC) and Cadence SoC Encounter are the two major place & route tools in industry and academic world.

1) IC Compiler

IC Compiler is a new generation of place & route tool developed by SYNOPSYS(to replace the previous generation—Astro). Astro runs independently of the layout, clock tree, and route, and has its limitations. The extended physical synthesis technology of IC Compiler breaks through this limitation and extends the physics to the whole place & route process. IC Compiler is a complete place & route design tool, which includes all the functions necessary for the next generation design, such as synthesis, place, wiring, timing and signal integrity optimization, testability design and yield optimization.

Compared with Astro, IC Compiler has faster running time, larger capacity, more intelligent multi-corner / multi-mode optimization, and improved predictability, which can significantly improve designers' productivity. At the same time, IC Compiler also launched a physical design that supports 32nm and 28nm technology. IC Compiler is becoming the ideal choice for leading Integrated Circuit Design Company in various applications. IC Compiler introduced the new technology for fast running mode, which reduced running time by 35% in the case of guaranteeing original quality.

2) SoC Encounter

Strictly speaking, SoC Encounter is not only a place & route tool, but also integrates a part of the functions of synthesis and static timing analysis. As a place & route tool, SoC Encounter supports the full chip design of 100 million gates on advanced technology of 28nm. In low power design, SoC Encounter can automatically divide different voltage domains in design process, and insert voltage regulator to balance various voltage. At the same time, clock tree, place and wiring is also optimized. SoC Encounter can also perform yield analysis in the process of RTL transformation to GDSII, and evaluate the impact of various place and route mechanisms, timing strategies, signal integrity and power consumption on yield, and ultimately get the best yield design plan.

1.6 CMOS device model

The CMOS device model mainly includes analysis model of large-signal, small-signal, noise and SPICE (Simulation Program with Integrated Circuit Emphasis) model for computer aided design, and these four kinds of models are discussed in detail in this section.

1.6.1 Large-signal model

In MOS transistor, drain current I_D is controlled by gate voltage, so we can consider that MOS transistor is a voltage-controlled current device. Large-signal model is shown in

Fig. 1.3, and the voltage-controlled current source represents drain current I_D. R_D and R_S respectively indicate contact resistor of drain and source. When I_D is small, the two resistors can be ignored. Two diodes represent PN junction formed by depletion layer between source-substrate and drain-substrate. Under normal working conditions, two PN junctions are reverse biased, and there are only leakage currents from the source / drain to substrate, which can be expressed by (1-1) and (1-2), where I_S is the reverse saturation current of PN junction. C_{GS}, C_{GD}, C_{GB}, C_{BS} and C_{BD} represent capacitors between each ports, which together determine the high frequency characteristics of MOS transistor.

$$I_{BS} = I_S \left(\exp\left(\frac{V_{BS}}{V_T}\right) - 1 \right) \tag{1-1}$$

$$I_{BD} = I_S \left(\exp\left(\frac{V_{BD}}{V_T}\right) - 1 \right) \tag{1-2}$$

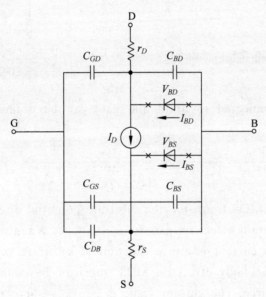

Fig. 1.3 Large-signal model

1.6.2 Small-signal model

We usually get DC operating point of circuit by analyzing large-signal model. This is a stable state. When there is a certain frequency of AC signal in the circuit, it is necessary to use AC small-signal model to analyze it. The AC small-signal model is based on its DC operating point. Because the analysis is response to small signal, the model can be linearized near operating point. The parameters in small-signal model are directly determined by current and voltage of DC operating point. The small-signal parameters of the same MOS transistor at different DC operating points are different. AC small-signal model reflects the MOS transistor response to a signal with a certain frequency, which is different from its DC characteristic. We can generate a small increment on DC bias point, and build a small-signal model by calculating the increment of other bias parameters caused by it. Because the drain current I_D is a function of gate-source voltage, and the low-frequency impedance between gate and source is very large, we can introduce a voltage-controlled current source $g_m V_{GS}$ to indicate, as shown in Fig. 1.4.

Due to channel-length modulation, the drain current I_D varies with the change of drain-source voltage. This effect can be expressed by a voltage-controlled current source. Moreover, because the value of the current source is linearly related to the voltage at both ends, we usually use an equivalent impedance r_o to simulate it, as shown in Fig. 1.5. The r_o is called intrinsic resistance, which determines the amplification gain of MOS transistor.

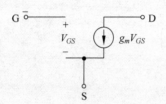

Fig. 1.4 Basic AC small-signal model

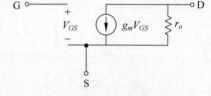

Fig. 1.5 Small-signal model with channel-length modulation effect

The resistance connected to drain and gate can be obtained by the following formula:

$$r_o = \frac{\partial V_{DS}}{\partial I_D} = \frac{1}{\partial I/\partial V_{DS}} = \frac{2}{\mu_n C_{ox} \frac{W}{L}(V_{GS} - V_{th})^2 \lambda} \approx \frac{1}{\lambda I_D} \qquad (1\text{-}3)$$

We know that in MOS transistor, the substrate potential directly affects threshold voltage, so it has a certain effect on gate-source voltage. When other ports maintain a constant voltage, drain current is also a function of substrate voltage. Therefore, it is possible to simulate the body effect on MOS transistor by connecting current source between drain and source. The current value is $g_{mb} V_{bs}$, where $g_{mb} = \partial I_D / \partial V_{bs}$, the small-signal model of MOS transistor considering body effect is shown in Fig. 1.6.

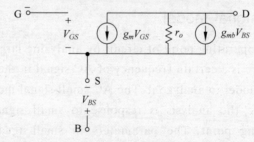

Fig. 1.6 Small-signal model of MOS transistor considering body effect

We have given the small-signal model considering channel-modulation effect and body effect. This model is suitable for lower operating frequency. As frequency increases, we have to consider capacitor effect between ports, the small-signal model of high frequency is demonstrated in Fig. 1.7.

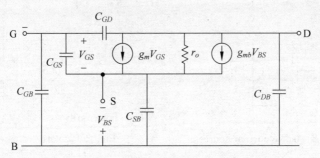

Fig. 1.7　Small-signal model of high frequency

Finally, we discuss the frequency characteristics of MOS transistor. We define that when MOS current gain falls to 1, the frequency is transit frequency f_T. For digital circuit, f_T characterizations the switching speed. For analog circuits, f_T represents the maximum operating frequency of MOS transistor. For a circuit, it usually operates at a much lower frequency than f_T, or less than $1/5 f_T$. The transit frequency of current MOS transistor can reach 40~60GHz, and the advanced MOS technology can reach 100GHz. Its expression is:

$$f_T = \frac{1}{2\pi} \frac{g_m}{C_{GS} + C_{GD} + C_{GB}} \tag{1-4}$$

f_T is the ratio of its transconductance to gate capacitor. When MOS works in saturation region, it is assumed that C_{GS} is far greater than $(C_{GD} + C_{GB})$, then formula (1-4) can be reduced to:

$$f_T = \frac{1}{2\pi} \frac{g_m}{C_{GS}} \tag{1-5}$$

1.6.3　Noise model

MOS noise is divided into three main categories: thermal noise, flick noise and shot noise. The following three kinds of noise models are discussed in a simple way.

1) Thermal noise

The MOS transistor has a channel resistance, so it can produce thermal noise as well as resistance. Its noise can be represented by a noise current source connected to drain and source end, as shown in Fig. 1.8. The channel equivalent resistance is:

$$R = 1/(\gamma g_m) \tag{1-6}$$

Where g_m is transconductance. γ depends on channel length, for long-channel MOS it is about 2/3, in short-channel devices γ will be much larger. So the thermal noise current of the MOS transistor can be obtained by formula (1-7).

$$\overline{i_n^2(f)} = 4\gamma k T g_m \tag{1-7}$$

The MOS thermal noise can also be expressed as a noise voltage source connected to gate, as shown in Fig. 1.9. The noise voltage can be expressed as:

$$\overline{V_n^2(f)} = 4\gamma kT/g_m \tag{1-8}$$

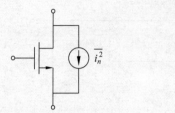
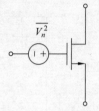

Fig. 1.8 Noise current source   Fig. 1.9 Noise voltage source

2) Flick noise

The flick noise is also called $1/f$ noise, which is caused by the pollution and defect of crystal surface. At the interface between the gate oxide and the silicon substrate in a MOSFET, since the silicon crystal reaches an end at this interface, many "dangling" bonds appear, giving rise to extra energy states. As charge carriers move at the interface, some are randomly trapped and later released by such energy states, introducing "flicker" noise in the drain current. In addition to trapping, several other mechanisms are believed to generate flicker noise. The flick noise can be expressed as a voltage source in series with gate, and its expression is as follows:

$$\overline{V_{n,flick}^2} = K/WLC_{ox}f \tag{1-9}$$

where K is a process-dependent constant on the order of $10^{-25} V^2 F$. The inverse dependence of (1-9) on W, L suggests that to reduce $1/f$ noise, the device area must be increased. It is therefore not surprising to see devices having areas of several hundred square microns in low-noise designs. PMOS devices exhibit less $1/f$ noise than NMOS transistors because the former carry the holes in a "buried channel," i.e., at some distance from the oxide-silicon interface, and hence trap and release the carriers to a less extent. Therefore, more PMOS transistors will be used in low noise design.

The total noise voltage of MOS transistor can be expressed by formula (1-10):

$$\overline{V_n^2} = \frac{K}{WLC_{ox}f} + 4KT\left(\frac{2}{3} \cdot \frac{1}{g_m}\right) \tag{1-10}$$

For a given device, in order to quantify the $1/f$ noise with thermal noise as a reference, the corresponding frequency is called corner frequency when the thermal noise is equal to $1/f$ noise. The intersection point serves as a measure of what part of the band is mostly corrupted by flicker noise. And the output current is determined as:

$$\frac{K}{WLC_{ox}f} = 4KT\left(\frac{2}{3} \cdot \frac{1}{g_m}\right) \tag{1-11}$$

that is:

$$f_c = \frac{K}{WLC_{ox}} \cdot \frac{3}{8KT} \cdot g_m \tag{1-12}$$

This result implies that f_C generally depends on the device area and transconductance.

3) Shot noise

Shot noise is caused by number variation of carriers in a conductor surface. It exists in diodes, MOS transistors and bipolar transistors. If the carrier density is very low and external electric field is high, the interaction between internal carriers can be neglected, and the noise current of unit bandwidth is obtained:

$$\overline{I_{ns}^2} = 2qI_D \tag{1-13}$$

where I_D is average current in transistor, q is the electronic charge 1.6×10^{-19} C, and for MOS transistor, because gate current is very weak, shot noise can almost be ignored.

1.6.4 Computer simulation model

The MOS transistor model discussed in the last section is convenient for manual calculation, but it ignores many second-order effects. Although a simple model for manual computation is essential, a more accurate model is needed for computer simulation.

SPICE(Simulation Program with Integrated Circuit Emphasis) is a general-purpose, open source analog electronic circuit simulator. It is a program used in integrated circuit design to check the integrity of circuit designs and to predict circuit behavior. SPICE was developed at the Electronics Research Laboratory of the University of California, Berkeley by Laurence Nagel. SPICE1 was first presented at a conference in 1973. SPICE1 was coded in FORTRAN and used nodal analysis to construct the circuit equations. Nodal analysis has limitations in representing inductors, floating voltage sources and the various forms of controlled sources. SPICE1 had relatively few circuit elements available and used a fixed-timestep transient analysis. The real popularity of SPICE started with SPICE2 in 1975. SPICE2, also coded in FORTRAN, was a much-improved program with more circuit elements, variable timestep transient analysis using either the trapezoidal (second order Adams-Moulton method) or the Gear integration method, equation formulation via modified nodal analysis(avoiding the limitations of nodal analysis), and an innovative FORTRAN-based memory allocation system. The last FORTRAN version of SPICE was 2G.6 in 1983. SPICE3 was developed in 1989. It is written in C, uses the same netlist syntax, and added X Window System plotting. Nowadays SPICE has been the most widely used computer simulation model standard for integrated circuit design.

Based on SPICE model, Berkeley developed BSIM1 model in 1984, which basically meets the needs of MOS transistor model above 0.8um process. BSIM1 model studies the modeling problem by using the multi-parameter curve fitting experiment. The model uses 60 parameters to describe MOS transistor's DC performance. In 1991, the improved BSIM2 model mainly considered the relationship between output resistance and Thermo Electron effect, source and drain parasitic resistance and inversion layer capacitor. And

BSIM2 model has 99 DC parameters. BSIM2 is a semiempirical model suitable for devices with channel length from long channel to about $0.2\mu m$. Because it also models the output resistance, BSIM2 is suitable for both digital and analog applications. In 1994, BSIM3 model was introduced. This model is easy to use and has only 40 DC parameters, but it can achieve good simulation performance for analog circuits.

In view of the complex physical effects of nanoscale process, BSIM4, developed in 2000 as the extension of BSIM3 model, addresses the MOSFET physical effects into sub-100nm regime. It is a physics-based, accurate, scalable, robustic and predictive MOSFET SPICE model for circuit simulation and CMOS technology development. Until now, BSIM4 has been used for the $0.13\mu m$, 90nm, 65nm, 45/40nm, 32/28nm, and 22/20nm technology nodes.

1) BSIM model

The model is based on MOS transistor's physical characteristics of small-geometric configuration, which takes into account the two states of weak inverse and strong inverse. Its main features include:

(1) The correlation between the carrier mobility and the vertical electric field.

(2) Charge sharing of source and drain.

(3) Channel narrowing.

(4) Subthreshold conduction.

(5) Nonuniform doping of ion implanted.

(6) Geometric configuration Association.

2) BSIM3V3 model

A significant improvement in BSIM3v3 model is the establishment of a unified I/V model. The model can describe current and output conductance characteristics from subthreshold to strong reverse region. This enhancement ensures that current conductivities and their derivatives remain continuous in all transition regions. And BSIM3v3 pays more attention to the important effects that the deep-submicron MOS transistor may have in its weak inversion region, mainly including the following aspects:

(1) vertical and lateral nonuniform doping effect.

(2) drain-induced barrier-lowering effect.

(3) normal and reverse short-channel effect.

(4) normal and reverse narrow width effect.

(5) field-dependent mobility and velocity saturation.

(6) channel-length modulation.

(7) impact ionization.

(8) polysilicon gate depletion.

3) BSIM4 model

As CMOS technology approaches more advanced levels, many novel physical effects appear, such as ①gate-induced drain leakage and direct-tunneling gate leakage; ②inversion layer quantization; ③finite charge layer effect; ④HF influence of MOSFET Parasitic; and ⑤ asymmetric source/drain resistance. Therefore in advanced technology like 0.13μm,90nm and more scaling technology, BSIM4 model including these new physical effects is widely adopted instead of BSIM3v3.

Compared with BIM3v3, BSIM4 made great improvement and function as below:

(1) Gate dielectric model: BSIM4 models the finite charge layer thickness(FCLT) in the channel effect by introducing an effective gate oxide capacitance C_{oxeff} in I-V and C-V models.

(2) Enhanced models for effective DC and AC channel length and width: Two additional fitting parameters XL and XW are introduced in BSIM4 to account for the offset in channel length and width due to the processing factors such as mask and etching.

(3) Enhanced Model for Nonuniform Lateral Doping due to Pocket(Halo) Implant: To reduce the short-channel effects(SCE), local high-dose implantation near the source/drain region edges have been employed. This is called lateral channel engineering or pocket(Halo) implantation, which causes higher doping concentration near the source/drain junctions than that in the middle of the channel. The pocket implantation will not only cause nonuniform lateral doping(NULD) but will also contribute to a nonuniform vertical doping, so the body effect will be influenced as well. A simple equation to consider the lateral nonuniform doping has been derived in BSIM3v3 (Cheng et al. (1997a)). It can model the Vth rolloff reasonably well as shown in Figure 6.7. An empirical term has been introduced in BSIM4 to improve the model accuracy with the consideration of the influence of NULD to the body effect.

(4) Improved Models for Short-channel Effects.

(5) Model for Narrow Width Effects.

(6) Channel charge model: BSIM4 uses a unified expression for the channel charge Q_{ch} from the strong inversion and subthreshold regions. The equation of the unified charge model in BSIM4 is similar to that in BSIM3v3, but additional improvements have been introduced to enhance the model accuracy in the transition region from the subthreshold to strong inversion.

(7) Mobility model: BSIM4 provides two mobility models that have been used in BSIM3v3. In addition, it also introduces a new mobility model based on the universal model given in Eq.(1-14).

$$\mu_{eff} = \frac{\mu_0}{1+(E_{eff}/E_0)^v} \tag{1-14}$$

where μ_0 is the low-field mobility, E_0 is called critical electric field, v is a constant, the value of which depends on the device type and technology. E_{eff} is an effective field defined empirically by

$$E_{eff} = \frac{Q_B + Q_{INV/2}}{\varepsilon_{si}} \tag{1-15}$$

where Q_B and Q_{INV} are the charge density in the bulk and in the channel, respectively.

(8) SOURCE/DRAIN RESISTANCE MODEL: in addition to the resistance model in BSIM3v3, which embeds the source/drain resistance in the I-V equation and assumes that series resistance at the source side equals the one at the drain side, BSIM4 introduces an asymmetric source/drain resistance model, which allows that the bias-dependent resistances at the source and the drain do not have to be equal and that they are physically connected between the external and the internal source/drain nodes. This asymmetric external source/drain resistance model is needed in simulating the high-frequency small-signal AC and noise behavior.

(9) Gate tunneling current model: In BSIM4, the gate tunneling current is modeled by several different components, the tunneling current between the gate and the substrate (I_{GB}), the current between the gate and the channel (I_{GC}), and the currents between the gate and the source/drain diffusion regions (I_{GS} and I_{GD}). I_{GC} can be further partitioned between the source and the drain, that is, $I_{GC} = I_{GCS} + I_{GCD}$.

(10) Substrate current models: The modeling of substrate current is important in today's MOSFETs, especially for analog circuit design. In addition to the junctions diode current and gate-to-body tunneling current, two other current components dominate the contribution of the total substrate current in a MOSFET: one is the substrate current due to impact ionization of the channel current and the other is the gate-induced drain leakage (GIDL) current.

(11) High-speed (Non-Quasi-Static) model: As the circuit clock frequency gets faster and faster, the need for an accurate prediction of device/circuit behavior near the cutoff frequency or under very rapid transient operations increases. In BSIM4, a charge-deficit non-quasi-static (NQS) model, based on the one in BSIM3v3 but with many improvements, is included.

(12) Layout-dependent parasitic model: BSIM4 considers the layout geometry-dependent parasitic model, which can calculate the parasitic geometry information according to the layout of the devices, such as isolated, shared, or merged source/drain, and multi-finger, and so on. Depending on the layout difference, many new model parameters have been introduced to select the model for the corresponding parasitic geometry calculation.

1.7 Technical words and phrases

1.7.1 Terminology

Integrated Circuit	集成电路
semiconductor	半导体
Complementary Metal-Oxide-Semiconductor(CMOS)	互补金属氧化物半导体
VLSI(Very Large Scale Integrated circuits)	超大规模集成电路

续表

Hardware Description Language(HDL)	硬件描述语言
Process Design Kit(PDK)	设计工具包
Electronic Design Automatic(EDA)	电子设计自动化
Design Rule Check(DRC)	设计规则检查
Layout Versus Schematic(LVS)	版图电路图一致性检查
Printed Circuit Board(PCB)	印刷电路板
Register Tranfer Level(RTL)	寄存器传输级
Static Timing Analysis(STA)	静态时序分析
Place and route	布局布线
Body effect	体效应
Thermal noise	热噪声
Flick noise	闪烁噪声
Shot noise	散粒噪声
Simulation Program with Integrated Circuit Emphasis(SPICE)	集成电路仿真程序
Berkeley Short-channel IGFET Model(BSIM)	伯克利短沟道绝缘栅场效应晶体管模型
channel-length modulation	沟道长度调制效应
Gate-Induced Drain Leakage(GIDL)	栅致漏极泄漏

1.7.2 Note to the text

(1) The silicon material is created by design and process, and integrates information collection, processing, operation, transmission and storage functions into a single chips, which has become the foundation of information technology in human society.

原始硅材料经过设计和工艺加工创造，将信息采集、处理、运算、传输和存储等功能集成并固化在硅芯片上，成为了人类社会信息化的基础。

(2) The biggest difference between CMOS analog integrated circuit design and traditional discrete-component analog circuit design is that all active and passive devices are made on the same substrate, and the size is extremely small, which can not be verified by circuit board. Therefore, the designer must use the method of computer simulation to verify the circuit performance.

CMOS模拟集成电路设计与传统分立元件模拟电路设计最大的不同在于，所有的有源和无源器件都是制作在同一衬底上，尺寸极其微小，无法再用电路板进行设计验证。因此，设计者必须采用计算机仿真和模拟的方法来验证电路性能。

(3) Due to the complexity and diversity of CMOS analog integrated circuits, there are no EDA vendors that can fully solve the design automation of CMOS analog integrated circuits, so all the analog circuits are still designed by hand.

由于CMOS模拟集成电路的复杂性和多样性，目前还没有EDA厂商能够提供完全解决 CMOS 模拟集成电路设计自动化的工具，因此所有的模拟电路基本上仍然通过手工设计来完成。

(4) In design process, the effects of design rules, matching, noise, crosstalk, parasitism on the performance and manufacturability of the circuit are considered. Although many advanced full custom aided design methods have emerged, it is still impossible to guarantee the comprehensiveness of manual design for layout and various effects.

在设计过程中需要考虑设计规则、匹配性、噪声、串扰和寄生效应等对电路性能和可制造性的影响。虽然现在出现了许多先进的全定制辅助设计方法，但仍然无法保证手工设计对版图布局和各种效应的考虑全面性。

(5) The physical verification phase will solve the above two kinds of validation problems through the design rule check(DRC) and the comparison between the layout netlist and the circuit schematic(Layout Versus Schematic, LVS).

物理验证阶段将通过设计规则检查(Desing Rule Cheek, DRC)和版图网表与电路原理图的比对(Layout Versus Schematic, VLS)来解决上述的两类验证问题。

(6) Compared with the traditional Virtuoso layout tools, the biggest highlight of Laker is the creative introduction of Schematic Driven Layout technology, that is, the transformation function of circuit schematic similar to printed circuit board(PCB) EDA tools.

相比传统的 Virtuoso 版图工具，Laker 最大的亮点在于创造性地引入电路图驱动版图技术(Schematic Driven Layout)，即实现了与印刷电路板 EDA 工具类似的电路图转换版图功能。

(7) The front-end design mainly includes function-structure analysis and design, RTL coding and simulation, synthesis, gate-level simulation and pre-layout static timing analysis. Back-end design consists of place and route, post-layout static timing analysis and post-simulation.

前端设计主要包括功能与结构分析设计、RTL 代码设计、RTL 级功能仿真、逻辑综合、综合后门级仿真、版图前静态时序分析；后端设计包括版图布局布线、版图后静态时序分析以及后仿真验证。

(8) Static Timing Analysis(STA) is an important step to ensure that the circuit meets the scheduled timing requirements by analyzing specific timing models and whether they violate the designer's given timing constraints.

静态时序分析是通过套用特定的时序模型(Timing Model)，针对电路分析其是否违反设计者给定的时序限制(Timing Constraint)，是保证电路满足预定时序要求的重要步骤。

(9) The flick noise is also called $1/f$ noise, which is caused by the pollution and defect of crystal surface.

闪烁噪声又称为 $1/f$ 噪声,它是由于晶体表面的污染及晶体缺陷所造成的。

(10) A significant improvement in BSIM3v3 model is the establishment of a unified I/V model. The model can describe current and output conductance characteristics from subthreshold to strong reverse region. This enhancement ensures that current conductivities and their derivatives remain continuous in all transition regions.

BSIM3v3 模型的一个显著提高是建立了统一的 I/V 模型,该模型可以描述从亚阈值直到强反型区的各个工作区域内,以及线性到饱和的各个状态下电流和输出电导特性,这一增强确保了电流电导率以及它们的导数在所有过渡区域内均能保持连续。

Part II

Part II

Chapter 2 Operational amplifier

Operational Amplifier (OPA), is named in 1947 by John R. Ragazzini, is used to represent a special type of amplifier. Through the rational allocation of external components, amplification, addition and subtraction, differential and integral mathematical operations can be completed for electrical signals. The actual OPA is usually used in the form of closed-loop negative feedback.

OPA is almost all over the analog and mixed-signal systems, from a simple amplifier to a complex Analog-to-Digital Converter (ADC). A large number of OPA assume different functions in different kinds of system: from bias voltage generation to signal sampling, hold, detection and filtering, OPA play an irreplaceable role in analog signal systems. With the progress of IC technology and decrease of supply voltage, all second-order effects are more and more obvious. The design of high-performance OPA is also facing more and more severe challenges.

2.1 Basis of OPA

OPA is a high gain amplifier. Fig. 2.1 gives its basic symbol. The Vip and Vin represent the same phase and the reverse input, respectively, and Vout represents the output end.

Fig. 2.2 is a typical two-stage differential OPA structure, it consists of five important parts: differential input stage, gain stage, output buffer, DC bias circuit and a phase compensation circuit (in order to drive load, OPA usually

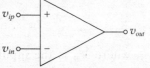

Fig. 2.1 OPA symbol

includes a output buffer). First of all, the differential input stage is usually a differential transconductance. The advantage of differential input is that it has a better common-mode rejection ratio (CMRR) than the single end. It converts input differential voltage signal to differential current and provides a differential to single conversion. A good transconductance should have good noise, offset, and linearity performance. The gain stage is the core of OPA, which plays a role of signal amplifying. In practical use, OPA often drives a low-impedance load, so an output buffer is needed to adjust the large output impedance to a small one to make output smoothly. The DC bias circuit provides a suitable DC operating point for transistors when OPA is working normally, so that the output AC signals can be loaded on DC operating point needed. The phase compensation circuit is used to stabilize OPA frequency characteristics. It shows that there is enough phase margin in frequency domain. And in time domain, it avoids the oscillation of output, and has a faster setup time.

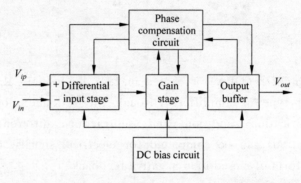

Fig. 2.2 The basic structure of two-stage differential OPA

2.1.1 Ideal OPA characteristics

The ideal OPA has infinite differential voltage gain, infinitely large input impedance and zero output impedance. The equivalent circuit of an ideal differential OPA is given in Fig. 2.3.

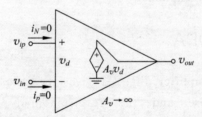

Fig. 2.3 The equivalent circuit of an ideal differential OPA

where V_d is the input differential voltage, that is, the difference between the two input terminals.

$$v_d = v_{ip} - v_{in} \tag{2-1}$$

Av is the differential voltage gain, and the output voltage is V_{out}

$$v_{out} = A_v(v_{ip} - v_{in}) \tag{2-2}$$

The differential voltage gain of ideal OPA is infinitely large, so its differential input voltage is very small and the differential input current is basically equal: $i_N \approx i_P \approx 0$.

2.1.2 Non-Ideal OPA characteristics

But in practice, the OPA parameters are far from ideal state. A typical two-stage OPA voltage gain is generally around 80~90dB(10 000~40 000), the input impedance is at mega level,and output impedance is about $10^5 \sim 10^6$ Ohom. Besides,OPA also has such non-ideal factors as noise, offset voltage, parasitic capacitor, gain nonlinearity, limited bandwidth, limited output swing and limited common mode rejection ratio. The equivalent circuit of non-ideal OPA is shown in Fig.2.4.

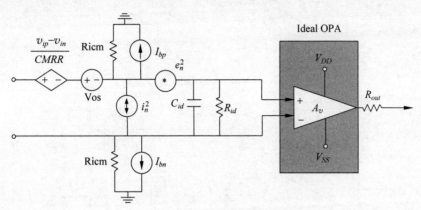

Fig. 2. 4 The equivalent circuit of non-ideal OPA

In Fig.2.4, R_{id} and C_{id} are the equivalent input resistor and capacitor, respectively, and Rout is the output equivalent resistor. The common mode input equivalent resistor is Ricm. Vos is the input offset voltage, which is defined as the input differential voltage when OPA output is zero. I_{bp} and I_{bn} are bias currents. CMRR is represented by voltage-controlled voltage source(vin/CMRR). This voltage source approximates the common mode influence. The noise source of OPA is represented by the equivalent noise voltage source e_n^2 and the equivalent noise current source i_n^2.

Fig.2.4 does not list all the non-ideal characteristics. In OPA design, the characteristics of limited gain-bandwidth product, finite slew rate and limited output swing are particularly important. The OPA output response is composed of a mixture of large signal and small signal characteristics. The set-up time of small signal is determined by poles and zeros in small-signal equivalent circuit, or the gain-bandwidth product and phase characteristic defined by its AC characteristic. But the set-up time of large signal is determined by output slew rate.

2.2 OPA structure

An OPA can be divided into a single-ended, a differential and a full-differential structure according to the input and output.

Fig.2.5 is a single-ended OPA structure. There is only one input and one output. It is very simple and has low power consumption, but the noise from power supply and ground can couple to output Vout, which seriously affect the output quality. Meanwhile its gain is small, with almost no use in practical design.

Fig.2.6 is a differential OPA structure with two differential inputs(v_{in} and v_{ip}) and

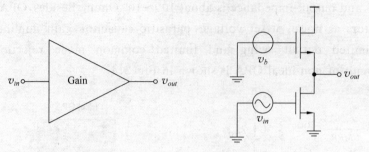

Fig. 2.5 single-end OPA structure

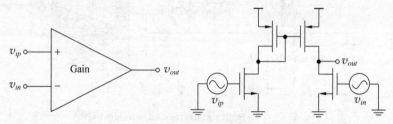

Fig. 2.6 Differential OPA structure

an output Vout, which has the advantages of simple structure. With same performance its power consumption is about two times larger than the single-ended one. The differential input will reduce common mode impact, so good common mode rejection performance is achieved. Usually it is used as a low-gain amplifier or a buffer circuit.

A fully-differential OPA structure is demonstrated in Fig. 2.7. It has two differential inputs(v_{ip} and v_{in}) and two differential output(voutp and voutn). Its output swing is two times large of differential structure. The full-differential circuit will suppress the noise influence from power supply and ground on output signal and reduce common mode effect. However, it needs a common mode feedback(CMFB) circuit to stabilize the common mode output.

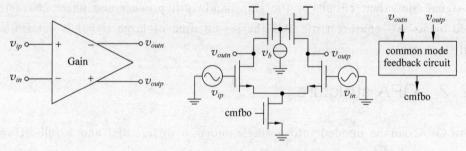

Fig. 2.7 Full-differential OPA structure

OPA can also be divided into single-stage and cascaded structures. A single-stage structure usually include telescopic cascode, folded cascode and gain-boosted. Cascode is used to improve the output resistance and DC gain. There are many poles in the multi-stage circuit, which may cause unstable in OPA use, so only the two-stage is designed usually.

2.2.1 Telescopic cascode

The fully differential telescopic cascode OPA is shown in Fig. 2.8, where M1 and M2 are input differential pairs, M3-M6 is cascode transistors, and M7, M8 and M0 are current source. The OPA gain is shown in equation (2-3).

$$A_v = g_{m1}[(g_{m3}r_{o3}r_{o1}) \| (g_{m5}r_{o5}r_{o7})] \quad (2-3)$$

Where g_{m1} is transconductance of input differential pairs, g_{m3} and g_{m5} are transconductance of cascode transistors M3 and M5, respecntively. r_{o1} represents channel resistor of M1 and M2, So r_{o3}, r_{o5} and r_{o7} for transistors M3, M5 and M7. Generally, the gain of fully-differential cascode structure OPA is easy to be designed to more than 80dB, but this is at the expense of reducing output swing and increasing poles.

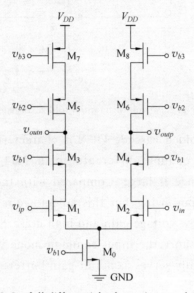

Fig. 2.8 full-differential telescopic cascode OPA

In Figure 2.8, its output swing is expressed as:

$$V_{sw,out} = 2[VDD - (V_{eff1} + V_{eff3} + V_{eff0} + | V_{eff5} | + | V_{eff7} |)] \quad (2-4)$$

The M_j overdrive voltage is represented by V_{effj} in the equation. Because there are 5 transistors stacking from power to ground, the output swing is limited. At the same time, another disadvantage of telescopic cascode is that it is difficult to achieve the same input common mode and output common mode voltage, that is, the short circuit between input and output can not be realized, which limits its application in negative feedback systems.

2.2.2 Folded cascode

The folded cascode is shown in Fig. 2.9, where M11 and M12 are input differential pairs, M3-M6 are cascode transistors, M1, M2, M7, M8 and M13 are current source. And its gain can be expressed as:

$$A_v = g_{m11}\{[g_{m3}r_{o11}(r_{o1} \| r_{o3})] \| (g_{m5}r_{o5}r_{o7})\} \quad (2-5)$$

g_{m11} is input transconductance, g_{m3} and g_{m5} are effective transconductance of cascode transistors M3 and M5, respectively. $r_{o1}, r_{o3}, r_{o5}, r_{o7}, r_{o11}$ are channel resistors of corresponding transistors.

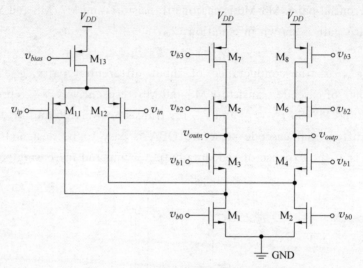

Fig. 2.9 full-differential folded cascade

The fully-differential folded cascode OPA is characterized by selection of voltage level, because the upper end of input differential pairs M11 and M12 are not cascaded, the input common mode range is large. Compared with the telescopic cascode, folded cascode OPA has larger output swing, but this advantage is obtained at the expense of larger power consumption, low voltage gain and high noise. Since the input and output of folded cascode OPA can be short, the input common mode voltage is easier to select. So the folded cascode OPA usually serves as a unit-gain buffer.

2.2.3 Gain-boosted

In Fig. 2.8 and 2.9, the telescopic and folded OPA uses a set of cascode transistors to increase output resistance and voltage. But if the supply voltage is not high enough, we want to further increase output resistance and voltage gain, the gain bootstrap is a feasible scheme.

Fig. 2.10 is two methods of increasing the equivalent output resistance. In Fig. 2.10(a) the equivalent output resistance Rout is improved by adding cascode transistor M2. If there is no cascode transistor M2, the equivalent output resistance is shown in equation (2-6), but after adding M2, it changes to equation (2-7).

$$R_{out} = r_{o1} \quad (2\text{-}6)$$

$$R_{out} = g_{m2} r_{o2} r_{o1} \quad (2\text{-}7)$$

where r_{o1} is the channel resistor of M1, g_{m2}, r_{o2} are transconductor and channel resistor of cascode transistor M2, respectively.

As shown in Fig. 2.10(b), adding an amplifier (the amplifier gain is A) between the

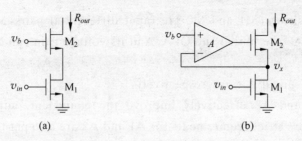

Fig. 2.10 Methods of increasing the equivalent output resistance

bias voltage and cascode transistor. And the M2 source and the amplifier is connected through a negative feedback mode, which makes the change of M2 drain voltage has less effect on node v_x. Due to the decrease of v_x voltage change, the output current is more stable through M1 output resistor, resulting in higher output resistance. The equivalent output resistance Rout can be obtained as shown in equation (2-8). And this structure is called gain-boosted.

$$R_{out} \approx A g_{m2} r_{o2} r_{o1} \quad (2-8)$$

As shown in equation(2-6), (2-7) and (2-8), The gain-boosted makes the equivalent output resistance and gain increase by about $A g_{m2} r_{o2}$ times. Its equivalent output resistance and gain are increased by A times compared to cascode structure. The purpose of improving output resistance can be achieved without increasing the number of cascode transistors. In general, the gain-boosted and cascode structures are used simultaneously, and the voltage gain above 110dB can be reached.

Fig. 2.11 shows a gain-boosted telescopic cascode OPA, where M0, M7 and M8 are

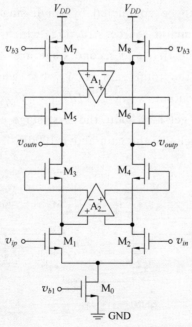

Fig. 2.11 Telescopic cascode OPA with gain-boosted

current source transistors, M1 and M2 are input differential pairs, M3-M6 are cascode transistors, A1 and A2 are similar sub-OPA. And its voltage gain can be expressed as:

$$A_v = g_{m1}[(g_{m3}r_{o3}r_{o1}A2) \| (g_{m5}r_{o5}r_{o7}A1)] \tag{2-9}$$

A1 and A2 are sub-OPA gains, respectively.

The OPA A1 and A2 effectively improve the equivalent output resistance and voltage gain, but more strict requirements for A1 and A2 are also put forward. Due to the negative feedback of A1 and A2, the conjugate zero and pole pairs (doublet) are introduced. The location of doublet may cause long set-up time for main OPA, so the frequency characteristics of A1 and A2 are regulated: the unit-gain bandwidth ω_g of A1 and A2 should be between the closed-loop bandwidth $\beta\omega_u$ of OPA and the non-dominant pole ω_{nd}, as shown in (2-10).

$$\beta\omega_u < \omega_g < \omega_{nd} \tag{2-10}$$

2.2.4 Two-stage

The gain of single-stage OPA is limited to the product of input transconductance and output impedance, while the cascode increases the gain but limits the output swing. Therefore the gain or output swing may not meet the requirements at the same design. To solve this problem, we can use two-stage structure to implement OPA circuit. The first stage provides high gain, and the second stage provides large output swing. Unlike the single-stage, the two-stage OPA can deal with voltage gain and output swing separately.

Fig. 2.12 demonstrates a two-stage OPA structure, where M0-M4 constitutes the first stage, M5-M8 constitutes the second, resistor Rc and capacitor Cc are used for frequency compensation. Each stage can be completed by single-stage amplifier, but the second stage is usually a simple common source structure, which can provide the maximum output swing. The gains of the first and second stage are $A_{v1} = g_{m1}r_{o1}r_{o3}$ and $A_{v2} = g_{m7}r_{o5}r_{o7}$, respectively. So the total gain is $A_{v1}A_{v2}$, which is almost equal to the cascode. But its differential output swing is $VDD - (|V_{eff5}| + |V_{eff7}|)$ that is much larger than its opponent. To get high gain, the first stage can be inserted into cascode transistors, and it is easy to achieve the gain above 100dB.

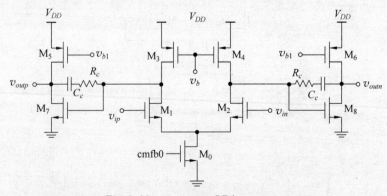

Fig. 2.12 two-stage OPA structure

Usually, we can also cascade more series OPA to achieve high gain. But in frequency characteristic, every stage at least introduces a pole in the open-loop transfer function. It is difficult to ensure the system stability in feedback system. So we have to add more complex circuits to compensate the system frequency characteristics.

2.3 Feedback

Because OPA need to be configured as negative feedback in the process of signal amplification, add and subtraction, integration and differential operation, the negative feedback and OPA are closely related and indivisible.

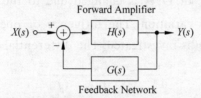

Fig. 2.13 Feedback system

Fig. 2.13 shows the basic structure of negative feedback system. The negative feedback system is usually composed of a feedforward amplifier, a feedback network and a summer, where $H(s)$ represents the transfer function of feedforward amplifier, and $G(s)$ represents the transfer function of the feedback network. It can be obtained from the structure:

$$[X(s) - Y(s)G(s)]H(s) = Y(s) \qquad (2\text{-}11)$$

that is

$$K(s) = \frac{Y(s)}{X(s)} = \frac{H(s)}{1 + G(s)H(s)} \qquad (2\text{-}12)$$

In equation (2-12), $H(s)$ is also called open-loop transfer function, and $K(s)$ represents the closed-loop transfer function. Usually the low frequency value of $H(s)$ is basically constant, called the low frequency small-signal gain, also known as the open-loop gain, which is represented by the A_v. And the feedback network is usually composed of passive devices, which is independent of frequency, and is represented by f. The equation (2-12) can be expressed as a form of (2-13).

$$A = \frac{A_v}{1 + A_v \cdot f} \qquad (2\text{-}13)$$

When a signal is propagating along a loop consisting of a feedforward amplifier, a feedback network, and a summer, the total gain of the signal is $A_v \times f \times (-1)$. Its absolute value is called the loop gain, which is recorded as T

$$T = A_v \cdot f \qquad (2\text{-}14)$$

When the open-loop gain tends to infinity, the closed-loop gain A of the ideal amplifier is

$$A_{ideal} = \lim_{A_v \to \infty} A = \frac{1}{f} \qquad (2\text{-}15)$$

It is known from equation (2-15) that when the open-loop gain tends to infinity, the closed-loop gain A is determined by the feedback network, which is independent of the open loop gain.

After adding negative feedback, OPA's characteristics will be improved in many ways, mainly in three aspects: gain desensitization, nonlinearity reduction and bandwidth modification.

2.3.1 Gain desensitization

The open-loop gain A_v of OPA is changed due to the factors such as integrated circuit technology and load variation. The influence of open-loop gain after negative feedback on closed-loop gain is investigated. The differential of equation (2-13) is

$$\frac{dA}{dA_v} = \frac{1}{(1 + A_v \cdot f)^2} \qquad (2\text{-}16)$$

put (2-14) into (2-16):

$$\frac{dA}{A} = \frac{1}{1+T} \frac{dA_v}{A_v} \qquad (2\text{-}17)$$

dA/A represents the relative change of the closed-loop gain, while dA_v/A_v represents the relative change of the open-loop gain. The equation (2-17) shows that the change of the loop gain is reduced $(1+T)$ times relative to it of the open-loop gain. The greater the loop gain T of a closed-loop amplifier is, the less sensitive the closed-loop gain A is to the change of open-loop gain.

2.3.2 Nonlinearity reduction

The linearity is a very important parameter of OPA. It is the linearity of the output signal to change with the input signal. If the gain varies with the change of input signal amplitude, the circuit will introduce a nonlinear distortion for output signal. Fig. 2.14 gives an OPA input and output characteristic curve. It can be seen that the gain is not a constant quantity, but changes with the amplitude of input signal. The larger the signal amplitude, the smaller the gain of the circuit.

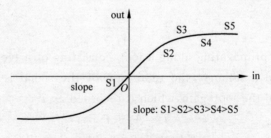

Fig. 2.14 Output versus input

If configured as negative feedback amplifier, the gain can be expressed as:

$$A = \frac{a_0}{1+T} \quad (2\text{-}18)$$

From (2-18), we can see that after adding negative feedback, the sensitivity of closed-loop amplifier is reduced, which makes the linearity greatly improved.

2.3.3 Bandwidth modification

Assuming a single-stage amplifier, its transfer function can be expressed:

$$H(s) = \frac{A}{1-s/p} \quad (2\text{-}19)$$

where P is the pole, that is, the -3dB bandwidth of amplifier, and its gain bandwidth is Ap.

The feedback network is usually realized by passive devices. After the negative feedback technology is applied, the feedback coefficient is usually a constant f. The transfer function of closed-loop amplifier is

$$A(s) = \frac{a(s)}{1+a(s)f} \quad (2\text{-}20)$$

put (2-19) into (2-20)

$$A(s) = \frac{a_0}{1+fa_0} \frac{1}{1-s/[(1+a_0 f)p]} = \frac{a(s)}{1+a(s)f} \frac{1}{1-s/[(1+T)p]} \quad (2\text{-}21)$$

It can be seen that the closed-loop amplifier is also a single-pole system with a bandwidth of $|(1+a_0 f)p|$. Its bandwidth is $|1+a_0 f|$ times larger than the open-loop one. But considering that the small-signal gain is $1/|1+a_0 f|$ that of the low-frequency signal, negative feedback will not change gain bandwidth product of the single-pole system. Even if the amplifier is a multi-pole system, as long as the other zero and poles are far from the dominant pole, the above conclusion is also established.

2.4 Parameters

1. DC-Gain(open-loop gain)

DC gain is defined as the ratio of the output voltage to the input voltage. The parameters are measured without load. In ideal case, the DC gain should be infinitely large; in fact, it is much less than the ideal condition. In operating, the DC gain difference between different OPA is up to 30%. Therefore, it is better to use OPA as a closed-loop system, when the open-loop gain only determines the accuracy of feedback system. The DC gain of a single-stage amplifier is the product of effective transconductance and output resistance, while in multi-stage amplifiers it is the product of gain of all single-stage amplifiers.

The ideal OPA should have an infinite open-loop gain, but the actual OPA limited gain usually brings a certain closed-loop gain error. We first assume that the open-loop gain of an OPA is 60dB, and is applied to the feedback amplifier in Fig. 2.15. Its gain can be obtained.

$$\frac{V_{out}}{V_{pos}} = \frac{1}{\frac{R_1}{R_1+R_f} + \frac{1}{A_{vol}}} \tag{2-22}$$

When the open-loop gain is close to infinity, the ideal gain can be obtained:

$$\frac{V_{out}}{V_{pos}} = \frac{R_f}{R_1} + 1 \tag{2-23}$$

When the open-loop gain is 60dB, the actual gain is as follows:

$$Gain = \frac{1}{\frac{R_1}{R_1+R_f} + \frac{1}{A_{vol}}} = 91.73 \tag{2-24}$$

The error is up to 9%, so the low open-gain operation is not suitable for high resolution amplification design.

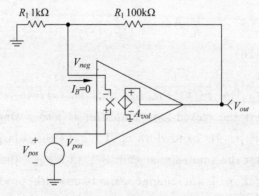

Fig. 2.15 Feedback amplifier

2. Phase Margin(PM)

Phase margin is an important parameter to measure the stability of OPA frequency characteristic. As shown in Fig. 2.16, it is defined as the difference between the phase point and the −180 degree phase point when the gain curve crosses the zero on bode plot. In theory, when phase margin is greater than 45 degrees, OPA is in a stable state. But in actual design, it is usually required OPA with a phase margin of more than 60 degrees.

3. Unit gain BandWidth(UGW)

The unit gain bandwidth defines as the small-signal bandwidth when OPA open-loop gain reduces to 1. Because OPA is usually used as closed-loop, we can use the Gain-BandWidth product(GBW) to express the relationship between the closed-loop gain and the small-signal bandwidth. That is, when closed-loop gain is 1, GBW is equal to UGW; in other words, GBW determines the maximum operating frequency OPA can handle. For example when the OPA closed-loop gain is 10, OPA's small-signal bandwidth decreases to 1/10 of UGW.

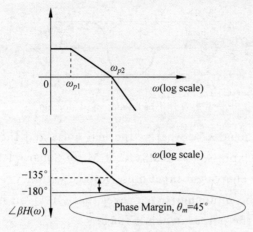

Fig. 2.16 Phase margin

4. Slew Rate(SR)

Slew rate indicates the responsiveness of an OPA to input signal change, and it is the parameter to measure its operation speed in the action of a large signal. Only when the absolute slope value of input signal is less than slew rate, the output can change linearly. In a simple sense, when OPA inputs a step signal, the maximum output speed(slew rate) is expressed as $SR = \mathrm{d}V/\mathrm{d}t\,|_{\max}$, its unit is V/us, as Fig. 2.17 shows.

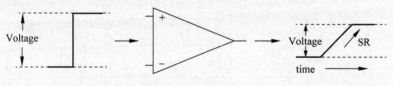

Fig. 2.17 Slew rate

Slew rate is an important parameter of the switched-capacitor circuit (integrator, sampling and holding circuit), which characterizes the ability of an OPA to receive the fast charging signal.

5. Input and Output Voltage Range

The two input terminals of OPA have a certain input swing limit, which are caused by the input stage design. In the prescribed work and load conditions, the output voltage range is defined as the extent to which the output can be driven to close to the positive or negative power rail. The output voltage range also depends on the output stage design and drive current in the test environment. When the amplitudes of OPA input(output) signal exceed the input(output) swing, the output signal will have the peak-clipping distortion, as shown in Fig. 2.18. The direct effect of input and output overload is to produce nonlinear distortion at the output end and affect the signal quality.

6. Equivalent input noise voltage

The equivalent input noise voltage is defined as: the output noise voltage is divided by the gain and equivalent to the input end, whose unit is $n\mathrm{V}/\sqrt{\mathrm{Hz}}$ or V^2/Hz. The

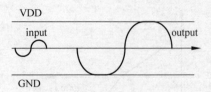

Fig. 2.18　peak-clipping distortion

equivalent input noise includes two parts: thermal noise and flick noise. The equivalent input noise can also be expressed as RMS(Root Mean Square, RMS) with unit(μV). Its value is to integrate the equivalent input noise voltage within a certain signal bandwidth. The noise determines the minimum signal level that OPA can handle and, to a certain extent, also determines the output dynamic range.

The equivalent input noise voltage of an OPA is mainly derived from the input differential transistors. The tail-current transistors and load transistors' noise need to be divided into the corresponding gain equivalent to the input port, so their contribution to the equivalent input noise voltage is less. In rough calculation, we can only calculate the noise voltage of input differential transistors as the equivalent input noise voltage. As an example, the equivalent input noise voltage of two-stage amplifier in Fig. 2.19 can be roughly calculated:

$$\overline{V_{n,in}^2} = \frac{16}{3}KT \cdot \frac{1}{g_{m.PM1}} + \frac{2K_P}{C_{ox}(WL)_{PM1}f} \tag{2-25}$$

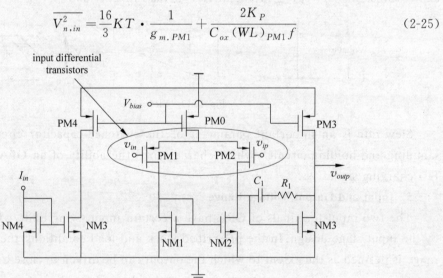

Fig. 2.19　The equivalent input noise voltage in two-stage OPA

7. Offest Voltage(Vos)

Ideally, when the voltage of two input terminals is equal, the output voltage of OPA is zero. But in practice, even if the voltage of the two input is the same, OPA also has a minimal voltage output. The input offset voltage is defined as the maximum voltage difference of the two input terminals when the output voltage is zero on the condition that the OPA operates in a linear area in a closed-loop circuit. The input offset voltage is usually defined at room temperature, and its unit is μV, as shown in Fig. 2.20.

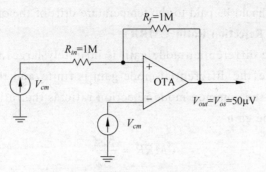

Fig. 2.20 Offset voltage

The offset voltage is derived from the mismatch between input differential transistors, as shown in Fig. 2.21. The mismatch is unavoidable because of the limitation of semiconductor technology. The matching degree of the differential transistors is proportional to the square root of transistor size in a certain range. When the transistor size is increased to a certain extent, continuous increase in area can not improve the matching. Increasing the area of the input transistors also means increasing the cost of chip. Usually a common method is to measure Vos after manufacture, and then by trimming to reduce the offset voltage.

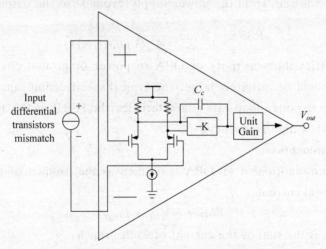

Fig. 2.21 Mismatch between input differential transistors

Though the offset voltage is usually defined at fixed temperature at room temperature, in practice, the input offset voltage always changes with temperature, which is directly related to the temperature characteristic of transistors. When the temperature changes, the input offset voltage temperature drift is defined as:

$$\frac{\Delta V_{os}}{\Delta T} = \frac{V_{os}(T_1) - V_{os}(25°)}{T_1 - 25°} \qquad (2\text{-}26)$$

At the same time, we should also pay attention to the long-term drift of offset voltage, whose unit is μV/1000hours or μV/month. Therefore, in design of wide range temperature OPA, special attention should be paid to the temperature drift of the offset voltage.

8. Common Mode Rejection Ratio(CMRR)

In ideal OPA, the differential-mode gain is infinitely large, and the common-mode gain is zero. In practice, the differential-mode gain is finite, and the common-mode gain is not zero. We define the common mode rejection ratio as the ratio of differential-mode gain and common-mode gain

$$CMRR = \frac{A_{dm}}{A_{cm}} \tag{2-27}$$

The more common way is to represent CMRR as a logarithmic form:

$$CMRR(\text{dB}) = 20\log_{10}\left(\frac{A_{dm}}{A_{cm}}\right) \tag{2-28}$$

The CMRR represents the sensitivity of an OPA to the change of common-mode voltage at the two input terminals. The CMRR of single power supply OPA is from 45dB to 90dB. Usually, when the OPA is used in a circuit when the input common-mode voltage changes with input signal, this parameter can not be ignored.

9. Power Supply Rejection Ratio(PSRR)

The power supply rejection ratio(PSRR) is defined as the gain from the input to the output divided by the gain from the power supply(ground) to the output.

$$PSRR = 20\log_{10}\frac{A_v(out)}{A_v(VDD,GND)} \tag{2-29}$$

PSRR quantifies the sensitivity of OPA to power or ground change. In the ideal case, the PSRR should be infinitely large. The typical specification range of the PSRR is 60dB to 100dB. As the open-loop gain characteristics, the PSRR of DC and low frequency is higher than that at high frequency.

10. Power consumption

The power consumption of an OPA is defined as the product of the power supply voltage and the total current:

$$Power = V_{DD} \cdot I_{total} \tag{2-30}$$

In Fig. 2.22, I_{total} is the sum of the current of each branch:

$$I_{total} = I_b + I_1 + I_{2a} + I_{2b} \tag{2-31}$$

In low power design, Power consumption is the most important design parameter. Power consumption can be seen as the source of OPA design. The total current is acquired through power consumption, and the branch currents are redistributed. According to the formula of "square-law", the width to length ratio of each transistor can be obtained.

11. Dynamic Range(DR)

When OPA is not distorted, DR is expressed as the ratio of the maximum output signal power to the noise floor power.

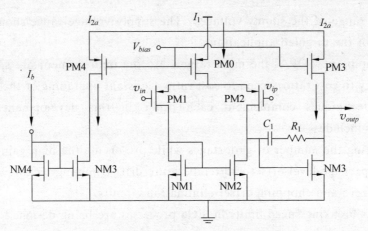

Fig. 2.22 Each branch current in an OPA

$$DR = 10\log_{10}\frac{P_{V_{out}}}{P_{Noise\ floor}} \qquad (2\text{-}32)$$

The unit is dB. DR is the most important parameter measuring the linear amplification of an OPA.

2.5 Industrial requirement of OPA design

A successful OPA product must conform to the industry standards and customer expectations by the parameters below:

(1) reliability performance: the OPA must be proved by the various tests(life test consisting of at least 168 hours at 150℃, set of the mechanical, humidity, etc. tests);

(2) electrostatic discharge protection: before released, the OPA must pass the test of three ESD models(HBM: Human Body Model; MM: Machine Model; CDM: Charged Device Model);

(3) package and pin assignment;

(4) operating temperature range(commercial: 0℃ to 70℃, industrial: −40℃ to 125℃, military: −55℃ to 125℃).

(5) absence of oscillations while operating with any gain.

(6) tolerance to at least 100 pF of the load capacitance and to a small or nonlinear load resistance.

(7) rail-to-rail output stage: The rail-to-rail output stage initially became a feature of the low-voltage OPA in order to improve the output operating range but developed into the customer expectation even for the high-voltage amplifiers as it simplifies application designs. Recently the rail-to-rail input stage capability is also developing in a customer expectation to any OPA by the same reason.

(8) absence of strange effects like output inversion when the input signal moves out of the specified range, Iq spike during output saturation, etc.

(9) wide range of the supply voltages: The supply voltage range should exceed the supply limits of the targeted applications.

In developing new OPAs the main targets are the improvement of: speed to power ratio; accuracy to cost ratio; speed to cost ratio. Constant upgrading of the processes and new circuit techniques complement each other in this development. The circuit enhancements include:

(1) reducing the number of gain stages while preserving the high gain;

(2) new package-level offset and temperature drift trimming techniques;

(3) auto-zero and chopping offset elimination circuits.

The OPAs breaking speed limits in SiGe processes are being designed for the signal processing in wireless and video applications. And for wearable device application, the very low supply voltage(<1V) designs are required.

2.6 Technical words and phrases

2.6.1 Terminology

Operational amplifier(OPA)	运算放大器(运放)
analog-to-digital converter(ADC)	模数转换器
common-mode rejection ratio(CMRR)	共模抑制比
AC signals	交流信号
frequency domain	频域
time domain	时域
fully-differential OPA	全差分运算放大器
telescopic cascode	套筒式共源共栅
folded cascode	折叠式共源共栅
gain-boosted	增益自举
doublet	零、极点对
feedforward amplifier	前馈放大器
DC-Gain(open-loop gain)	直流增益(开环增益)
Phase Margin(PM)	相位裕度
Unit gain BandWidth(UGW)	单位增益带宽
Gain-BandWidth product(GBW)	增益带宽积
Slew Rate(SR)	压摆率
Equivalent input noise voltage	等效输入噪声电压
Offest Voltage(Vos)	失调电压
trimming	修调
Power Supply Rejection Ratio(PSRR)	电源抑制比
Power consumption	功耗
Dynamic Range(DR)	动态范围
Human Body Model(HBM)	人体模型
Machine Model(MM)	机器模型
Charged Device Model(CDM)	带电器件模型

2.6.2　Note to the text

(1) A large number of OPA assume different functions in different kinds of system: from bias voltage generation to signal sampling, hold, detection and filtering, OPA play an irreplaceable role in analog signal systems.

大量运算放大器在电路系统中承担着不同的功能：从偏置电压产生到信号采样保持、检测和滤波，运算放大器在模拟信号系统中都扮演着不可替代的角色。

(2) The ideal OPA has infinite differential voltage gain, infinitely large input impedance and zero output impedance.

理想的运算放大器具有无限大的差模电压增益、无穷大的输入阻抗和为零的输出阻抗。

(3) The set-up time of small signal is determined by poles and zeros in small-signal equivalent circuit, or the gain-bandwidth product and phase characteristic defined by its AC characteristic. But the set-up time of large signal is determined by output slew rate.

运算放大器的输出响应由大信号特性和小信号特性混合构成，小信号建立时间完全由小信号等效电路的极点和零点的位置决定，或者由其交流特性定义的增益带宽积以及相位特性决定，而大信号的建立时间由输出压摆率决定。

(4) At the same time, another disadvantage of telescopic cascode is that it is difficult to achieve the same input common mode and output common mode voltage, that is, the short circuit between input and output can not be realized, which limits its application in negative feedback systems.

同时，套筒式共源共栅运放的另一个缺点是较难实现输入共模电压与输出共模电压的一致，即无法实现输入与输出短路，限制了其在负反馈系统的应用。

(5) Compared with the telescopic cascode, folded cascode OPA has larger output swing, but this advantage is obtained at the expense of larger power consumption, low voltage gain and high noise.

折叠式共源共栅运算放大器与套筒式结构相比，其输出摆幅较大，但这个优点是以较大的功耗、较低的电压增益、较低的极点频率和较高的噪声为代价获得的。

(6) But in frequency characteristic, every stage at least introduces a pole in the open-loop transfer function. It is difficult to ensure the system stability in feedback system. So we have to add more complex circuits to compensate the system frequency.

但是在频率特性中每一级在开环传输函数中至少引入一个极点，在反馈系统中使用这样的多级运放很难保证系统稳定，必须增加复杂的电路对系统进行频率补偿。

(7) After adding negative feedback, OPA's characteristics will be improved in many

ways, mainly in three aspects: gain desensitization, nonlinearity reduction and bandwidth modification.

放大器加入负反馈后,其特性在多方面会得到改善,主要表现为降低增益灵敏度、提高线性度和拓展带宽三个方面。

(8) Slew rate indicates the responsiveness of an OPA to input signal change, and it is the parameter to measure its operation speed in the action of a large signal.

压摆率表示运放对信号变化速度的反应能力,是衡量运放在大幅度信号作用时的工作速度的指标。

(9) The tail-current transistors and load transistors' noise need to be divided into the corresponding gain equivalent to the input port, so their contribution to the equivalent input noise voltage is less.

运放中的尾电流晶体管、负载晶体管噪声都需要除以相应的增益等效到输入端,因此对等效输入噪声电压贡献较小。

(10) The offset voltage is derived from the mismatch between input differential transistors. The mismatch is unavoidable because of the limitation of semiconductor technology.

运放输入失调电压来源于运放差分输入晶体管尺寸之间的失配。受工艺水平的限制,这个失配不可避免。

Chapter 3　Bandgap and LDO

Voltage sources in integrated circuits are usually divided into voltage reference sources and linear regulators. As a stable reference, the bandgap reference and low-dropout(LDO) linear regulator is very important in different modules of analog and mixed-signal integrated circuits. As the voltage and current source the output voltage of bandgap has characteristics independent of temperature. And LDO provides a stable, pure DC voltage reference for on-chip circuits as a basic power supply system.

3.1　Bandgap

As a very important module of analog and mixed-signal IC, the voltage reference source plays a very important role in various electronic systems. With the increasing performance demand for various electronic products, the requirements for voltage reference source are also increasing. The voltage source's output voltage and noise determines the performance of circuits and systems.

The bandgap is fully compatible with standard CMOS process and can work at low power supply voltage. In addition, it has low temperature drift, low noise and high power supply rejection ratio, which can meet the requirements of most electronic systems. With these advantages, the bandgap has been widely studied and applied. In CMOS bandgap, low power supply voltage, low power consumption, high precision and high PSRR are the future development direction.

3.1.1　Basic of bandgap

Many modules in integrated circuits require voltage sources and current sources independent of temperature, which often affects the function of these modules. So how can we get a constant voltage or current reference that has nothing to do with the temperature? We assume that there are two identical physical quantities in the circuit. These two physical quantities have opposite temperature coefficients. When the two physical quantities are added to a certain weight, the voltage reference of zero temperature coefficient can be obtained, as shown in Fig.3.1.

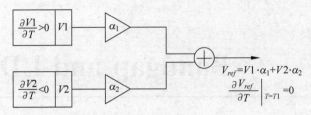

Fig. 3.1 Principle diagram of zero temperature coefficient

In Fig. 3.1, the voltage source $V1$ has a positive temperature coefficient $\left(\frac{\partial V1}{\partial T}>0\right)$, and the voltage source $V2$ has a negative temperature coefficient $\left(\frac{\partial V2}{\partial T}<0\right)$. We choose two weights of α_1 and α_2 to satisfy $\alpha_1 \cdot \frac{\partial V1}{\partial T} + \alpha_2 \cdot \frac{\partial V2}{\partial T} = 0$, then the voltage reference of zero temperature coefficient is obtained: $V_{ref} = V1 \cdot \alpha_1 + V2 \cdot \alpha_2$. The following task is how to get two voltage $V1$ and $V2$ with opposite temperature coefficients. In semiconductor technology, bipolar transistors can provide physical quantities of positive and negative temperature coefficient respectively. They are widely used in the design of bandgap reference. Recently, various literatures have also mentioned that the positive and negative temperature coefficients can be obtained by using MOS transistors working in subthreshold regions, but the accuracy of subthreshold region models needs to be investigated further. And the modern standard CMOS process provides the model of longitudinal PNP bipolar transistor, making the bipolar transistor still the first choice for bandgap reference.

1. Negative temperature coefficient voltage $\left(\frac{\partial V2}{\partial T}<0\right)$

For a bipolar transistor, the relationship between the collector current I_c and the base-emitter voltage V_{BE} is as follows

$$I_c = I_s \cdot \exp(V_{BE}/V_T) \tag{3-1}$$

$$V_{BE} = V_T \cdot \ln(I_C/I_s) \tag{3-2}$$

In equation (3-1) and (3-2), I_s is the saturation current of transistor, V_T is the thermal voltage, $V_T = kT/q$, K is the Boltzmann constant, and Q is electronic charge. The derivative of equation (3-2) for V_{BE} is

$$\frac{\partial V_{BE}}{\partial T} = \frac{\partial V_T}{\partial T} \ln \frac{I_c}{I_s} - \frac{V_T}{I_s} \frac{\partial I_s}{\partial T} \tag{3-3}$$

Obtained from the theory of semiconductor physics:

$$I_s = b \cdot T^{4+m} \exp \frac{-E_g}{kT} \tag{3-4}$$

The derivative of (3-4) for temperature:

$$\frac{V_T}{I_s} \frac{\partial I_s}{\partial T} = (4+m)\frac{V_T}{T} + \frac{E_g}{kT^2}V_T \tag{3-5}$$

from equation(3-3) and (3-5):

$$\frac{\partial V_{BE}}{\partial T} = \frac{V_{BE} - (4+m)V_T - E_g/q}{T} \tag{3-6}$$

where $m \approx 1.5$, When the substrate is silicon, $E_g = 1.12\text{eV}$. While $V_{BE} = 750\text{mV}$, $T = 300\text{K}$, $\frac{\partial V_{BE}}{\partial T} = -1.5\text{mV}/\text{°C}$.

It is known from equation (3-6) that the temperature coefficient $\frac{\partial V_{BE}}{\partial T}$ of V_{BE} itself is related to the temperature T. If the positive temperature coefficient is a temperature independent value, there will be errors in temperature compensation, resulting in a voltage reference that can only get a zero temperature coefficient at one temperature point.

2. Positive temperature coefficient voltage $\left(\frac{\partial V1}{\partial T} > 0\right)$

If two identical bipolar transistors are biased at different collector current, the difference between their base-emitter voltage is proportional to the absolute temperature, as shown in Fig.3.2.

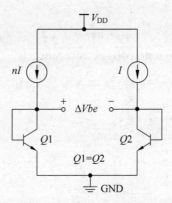

Fig. 3.2 Positive temperature coefficient voltage circuit

As shown in Fig.3.2, two identical bipolar transistors $Q1$ and $Q2$, which are biased at different collector current I_0 and nI_0. Ignoring their base current, there is:

$$\Delta V_{BE} = V_{BE1} - V_{BE2} = V_T \ln\frac{I_{c1}}{I_{s1}} - V_T \ln\frac{I_{c2}}{I_{s2}} = V_T \ln\frac{nI}{I_{s1}} - V_T \ln\frac{I}{I_{s2}} \tag{3-7}$$

There also is $I_{s1} = I_{s2} = I_s$, $I_{c1} = nI_{c2}$, then

$$\Delta V_{BE} = V_T \ln\frac{nI}{I_s} - V_T \ln\frac{I}{I_s} = V_T \ln n = \frac{kT}{q}\ln n \tag{3-8}$$

The derivative of (3-8) for temperature:

$$\frac{\partial \Delta V_{BE}}{\partial T} = \frac{k}{q}\ln n > 0 \tag{3-9}$$

We can see that there is a positive temperature coefficient in equation (3-9), which is independent of temperature T.

3. Zero temperature coefficient voltage reference $\left(\dfrac{\partial V_{REF}}{\partial T}=0\right)$

By using the voltage of positive and negative temperature coefficients obtained in the above two sections, a voltage reference V_{REF} independent of temperature can be obtained, as shown in Fig. 3.3, equation (3-10) can de derived:

$$V_{REF} = \alpha_1 \cdot \frac{kT}{q}\ln n + \alpha_2 \cdot V_{BE} \tag{3-10}$$

The following shows how to select α_1 and α_2, then get the zero temperature coefficient voltage V_{REF}. At room temperature (300K), there is a negative temperature coefficient voltage $\dfrac{\partial V_{BE}}{\partial T} = -1.5\text{mV/K}$, and positive temperature coefficient voltage is $\dfrac{\partial \Delta V_{BE}}{\partial T} = \dfrac{k}{q}\ln n = 0.087\text{mV/K} \cdot \ln n$

The derivative of (3-10) for temperature

$$\frac{\partial V_{REF}}{\partial T} = \alpha_1 \cdot \frac{k}{q}\ln n + \alpha_2 \cdot \frac{\partial V_{BE}}{\partial T} \tag{3-11}$$

assuming (3-11) equal to zero and $\alpha_2 = 1$, put the positive and negative temperature coefficient voltage in (3-11):

$$\alpha_1 \cdot \ln n = 17.2 \tag{3-12}$$

So the zero temperature coefficient voltage reference is

$$V_{REF} \approx 17.2\frac{kT}{q} + V_{BE} \approx 1.25\text{V} \tag{3-13}$$

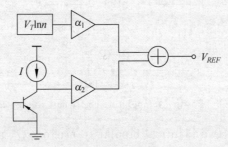

Fig. 3.3　Zero temperature coefficient voltage reference

4. Circuit of Zero temperature coefficient voltage reference

From the analysis of the previous section, the zero temperature coefficient voltage reference is obtained by adding the base-emitter voltage V_{BE} and $17.2 * kT/q$. The zero temperature coefficient voltage reference circuit is shown in Fig. 3.4. Suppose the voltage $V1 = V2$ in Fig. 3.4, then the left and right branches are equation (3-14) and (3-15), respectively.

$$V1 = V_{BE1} \tag{3-14}$$

$$V2 = V_{BE2} + IR \tag{3-15}$$

then

$$V_{BE1} = V_{BE2} + IR \tag{3-16}$$

so

$$IR = V_{BE1} - V_{BE2} = kT/q \cdot \ln n \tag{3-17}$$

put (3-17) into (3-15),

$$V2 = V_{BE2} + kT/q \cdot \ln n \tag{3-18}$$

Comparing (3-18) with (3-13), it is known that this circuit can obtain zero temperature coefficient voltage reference. The problem is how to make the voltage at both ends of the circuit equal in Fig. 3.4, that is, $V1 = V2$?

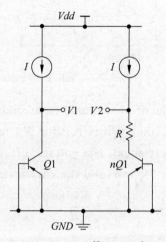

Figure 3.4 zero temperature coefficient voltage reference circuit

We know that when the ideal OPA operates normally, the voltage at the two input is approximately equal, so the following two circuits in Fig. 3.5 and 3.6 can be generated, making $V1 = V2$, respectively. There are two main kinds of circuit structure to complete adding, one is to add the two through an OPA. And its output is the voltage reference. The other is to generate a current proportional to absolute temperature (PTAT), which can be converted into a voltage through a resistor. This voltage naturally has a positive temperature coefficient, which is then added to the base-emitter voltage V_{BE}.

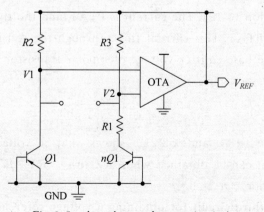

Fig. 3.5 the voltage reference circuit A

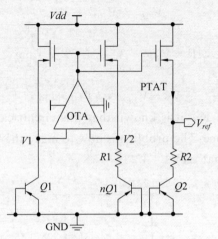

Fig. 3.6　the voltage reference circuit B

In Fig. 3.5, the input voltage of operational transconductance amplifier(OTA) is $V1$ and $V2$, and the output is V_{ref} that drives resistor $R2$ and $R3$. OTA makes the input voltage $V1$ and $V2$ approximately equal. The voltage difference between the base-emitter of the two bipolar transistors is $V_T \ln n$, and the current that flows through $R1$ is

$$I_2 = \frac{V_T \ln n}{R_1} \tag{3-19}$$

get V_{ref} is,

$$V_{ref} = V_{BE,nQ1} + \frac{I_2}{R_1} \cdot (R_1 + R_3) = V_{BE,nQ1} + \left(1 + \frac{R_3}{R_1}\right) V_T \ln n \tag{3-20}$$

Combination of (3-13) and (3-20) shows that at room temperature 300K, zero temperature coefficient voltage can be obtained: $V_{ref} \approx 1.25\text{V}$.

Fig. 3.6 is another circuit for obtaining a voltage reference. The principle of this circuit is to generate a current which is directly proportional to absolute temperature, and then convert it to voltage through resistor. Finally, the voltage is added to V_{BE} of bipolar transistor to get the voltage reference. As in Fig. 3.6, the current produced by the middle branch is still as equation (3-19). The current is PTAT, and the right mirror branch also produces a current of PTAT. This current flows through the resistor to form the PTAT voltage, and finally the base-emitter voltage of bipolar transistor $Q2$ is added to obtain the voltage reference

$$V_{ref} = V_{BE,Q2} + \frac{R_2}{R_1} V_T \ln n \tag{3-21}$$

The combination of (3-13) and (3-21) shows that a voltage reference of zero temperature coefficient can be obtained when $R2(\ln n)/R1 = 17.2$ is used. Then when $R2/R1 = 10$, we can choose $n = 6$.

Figure 3.7 is the third circuit for obtaining a voltage reference. The basic principle of this circuit is similar to that of second kinds. The difference is that two resistors

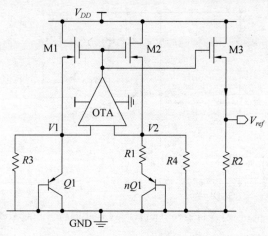

Fig. 3.7 the voltage reference circuit C

($R3 = R4 = R$) are added to the nodes $V1$ and $V2$ respectively, and the current flowing through the resistor is $I_R = V2/R4 = V_{BE1}/R$, So the current I_{M2} flows through the MOS transistor:

$$I_{M2} = I_R + I_{R1} = \frac{V_{BE1}}{R} + \frac{V_T \ln n}{R_1} \tag{3-22}$$

If the MOS transistors size $(W/L)_3 = (W/L)_2$, then there is $I_{M3} = I_{M2}$, and the voltage reference V_{ref} is obtained

$$V_{ref} = I_{M3} R_2 = \left(\frac{V_{BE1}}{R} + \frac{V_T \ln n}{R_1}\right) R_2 \tag{3-23}$$

It is known from equation (3-23) that the voltage reference of this structure can be obtained by adjusting the resistor value of $R2$. The first two structures can only get the reference voltage of 1.25V.

5. Offset in OTA

In Fig. 3.8, V_{os} is the offset voltage of OTA. The offset voltage makes the voltage reference error, and this error is also related to the temperature. Suppose that the OTA is ideal, so there is

$$V_{BE,Q1} - V_{os} \approx V_{BE,nQ1} + R_1 I_c \tag{3-24}$$

where I_c is the current flowing through resistor $R1$ and bipolar transistor $nQ1$, and the output voltage of OTA is

$$V_{ref} = V_{BE,nQ1} + (R_1 + R_3) I_c = V_{BE,nQ1} + (R_1 + R_3) I_c \tag{3-25}$$

combine (3-24) with (3-25):

$$V_{ref} = V_{BE,nQ1} + \left(1 + \frac{R_3}{R_1}\right)(V_T \ln n - V_{os}) \tag{3-26}$$

$$V_{ref} = V_{BE,nQ1} + \left(1 + \frac{R_3}{R_1}\right) V_T \ln n - \left(1 + \frac{R_3}{R_1}\right) V_{os} \tag{3-27}$$

According to equation (3-27), there is an error in the output voltage reference

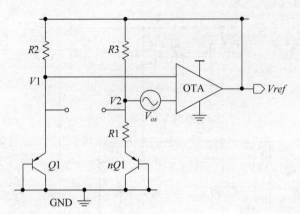

Fig. 3.8 Offset of OTA in Zero temperature coefficient voltage reference circuit

because of the offset voltage in OTA. This error is the result of $(1 + R3/R1)$ times of OTA offset. There are several circuits that can reduce the offset voltage:

(1) Increase the area of input transistors of an OTA, reduce the offset voltage by the design of the common-centroid layout.

(2) In high power supply, two bipolar transistors can be used in series, which makes ΔV_{BE} increase.

(3) Diffrenent current of two branches makes ΔV_{BE} increase from $\ln n$ to $\ln(MN)$, and reduces the ratio of V_{os} to V_{ref}, of which m and N are positive integers.

6. Start-up of bandgap

As shown in Fig. 3.6, the bandgap actually has two operating points, one is the working point when the circuit is normal, and the other is the zero current point. In power-on process, all the transistors in the circuit do not have current, and this state will be kept forever without no external interference. This is the problem of start-up.

In order to solve the start-up problem, an additional circuit is needed. The basic requirement of the start-up circuit is that after the power supply is stable, when circuit is in the "zero current" working state, the start-up circuit gives a stimulus to the internal circuit node, which forces it to get rid of the "zero current" working state and fall into the normal working mode. And when circuit is in normal operating mode, the start-up circuit stops working. The start-up circuit is in the right part in Fig. 3.9.

In Fig. 3.9, when the supply voltage is normally supplied and there is no current in the reference circuit, that is, PMOS transistor $PM4$ and $PM5$ have no current to pass through. While the node $Net1$ voltage is zero, PMOS transistor $PM3$ turns on, NMOS transistor $NM1$ cuts off. So the voltage of node $net3$ is $Vdd - 2V_{BE}$, which turns on NMOS transistor $NM2$ and makes node $net4$ connect to ground. Finally, the PMOS transistor $PM4$ and $PM5$ are on, and the voltage of node $Net1$ gradually increases to about $2V_{BE}$. The bandgap get to be operating normally.

Then in start-up circuit, the NMOS transistor $NM1$ is on, and $PM3$ is off, which makes the voltage of node $net3$ decrease gradually and $NM2$ is gradually cut off. After

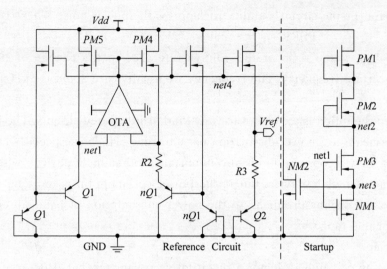

Fig. 3.9 Bandgap with start-up circuit

the normal operation of bandgap circuit, the two branches of start-up circuit stop working(NM2/PM3 is off).

3.1.2 Bandgap design

The performance of bandgap determines the precision that analog circuit can achieve, so its performance paremeter needs to be set according to the precision of functional module. Because the bandgap circuit mainly provides DC voltage(current), its low-frequency AC (< 10kHz), temperature and voltage characteristics are highly demanded. The main design difficulty is that it has good AC characteristics and stability while obtaining a lower temperature coefficient. The following is a brief introduction to its performance parameter.

1. Parameter

1) Temperature Coefficient(TC)

The temperature coefficient is a performance specification to measure the output voltage as a function of temperature. It is usually represented by ppm(parts per million). The equation is shown in (3-28).

$$\mathrm{TC(ppm/℃)} = \frac{V_{\max} - V_{\min}}{V_{\mathrm{mean}}(T_{\max} - T_{\min})} \times 10^6 \tag{3-28}$$

where $V_{\max}$ and $V_{\min}$ are the maximum and minimum values of voltage reference obtained at required temperature range. V_{mean} is the average, while $T_{\max}$ and $T_{\min}$ are the maximum and minimum temperature.

2) Power Supply Rejection Ratio(PSRR)

Power supply rejection ratio is a parameter to measure the ability of output voltage to suppress the variation of power supply voltage. Because the power supply voltage is not fixed during the normal operation, there are various kinds of noises. The greater the

PSRR, the stronger the circuit's ability to suppress the power noise. There is

$$\text{PSRR}|_{\text{Hz}} = -20\log_{10}(\partial V_{ref}/\partial V_{DD})(\text{dB}) \qquad (3\text{-}29)$$

$\partial V_{ref}/\partial V_{DD}$ is the ratio of the voltage reference to the change of power supply voltage at a certain frequency. The frequency we usually care about is 1kHz and 10kHz.

3) Power

Power consumption is a concern for any kind of integrated circuits. The lower power consumption means less power dissipation per unit time, which is especially important for IC packaging. Power consumption is also important for handheld devices, which is related to the lifetime of mobile device. But if the power consumption is too demanding, there may be noise and driving problems, so the power consumption of bandgap circuit should be in a reasonable range.

4) Start-up

The start-up of bandgap is not a quantitative parameter, but it determines whether its function is normal. Start-up problems are not observed in the usual simulation of transient, DC, and parameter scanning. Since there are two DC operating points in bandgap, if we want bandgap to leave the zero current state and enter normal working mode, we need to add the necessary "stimulus" to start the circuit. The normal operation of bandgap is usually confirmed by adding 0 to V_{DD} slope voltage signals on power supply.

2. Circuit design

A complete circuit of a bandgap is shown in Fig. 3.10.

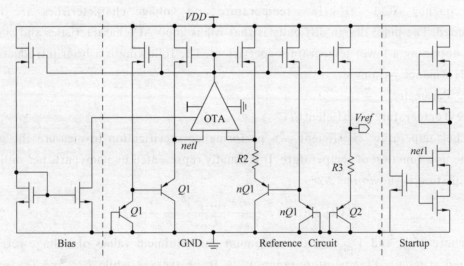

Fig. 3.10 Bandgap circuit

The bandgap is mainly divided into three parts, from left to right, followed by voltage bias circuit(Bias), voltage reference generator(Reference Circuit) and start-up circuit(Start-up). The voltage bias circuit is used to generate bias voltage when OTA works normally. From the Fig. 3.10, we can see that voltage bias comes from voltage reference generator. The voltage reference generator is used to generate the required voltage reference(1.25V). We usually use the bipolar transistors in series to obtain the positive temperature coefficient voltage, which is helpful to reduce the influence of OTA offset to reference voltage.

As the positive temperature coefficient voltage is obtained by using the bipolar transistors in series, the current of resistor $R2$ is I_2 in Fig.3.10.

$$I_2 = \frac{2V_T \ln n}{R_2} \quad (3\text{-}30)$$

The resulting voltage reference V_{ref}

$$V_{ref} = V_{BE,Q2} + \frac{2R_3}{R_4} V_T \ln n \quad (3\text{-}31)$$

$$\frac{\partial V_{REF}}{\partial T} = \frac{\partial V_{BE,Q2}}{\partial T} + \frac{2R_3}{R_4} \cdot \frac{k}{q} \ln n \quad (3\text{-}32)$$

put $\dfrac{\partial V_{BE}}{\partial T} = -1.5\text{mV/K}, \dfrac{\partial \Delta V_{BE}}{\partial T} = \dfrac{k}{q} \ln n = 0.087\text{mV/K} \cdot \ln n$ into (3-32)

$$\frac{R_3}{R_4} \cdot \ln n = 8.6 \quad (3\text{-}33)$$

When we choose $n = 7, R_3 \approx 5R_4$; And $R_3 = 5\text{k}\Omega, R_4 = 25\text{k}\Omega$.

For OTA in a bandgap, the more important performance parameters are DC gain, gain bandwidth product, phase margin, and power supply rejection ratio. The OTA circuit is shown in Fig.3.11.

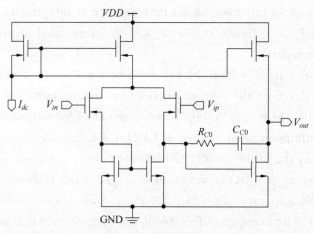

Fig. 3.11 OTA circuit

Because the output voltage accuracy and PSRR of bandgap is related to the DC gain of OTA, the two-stage structure is selected to increase its DC gain. The first stage uses a simple five-transistor structure to gain medium gain, and the second stage uses common source to provide a certain gain while providing a larger output swing. The capacitor C_{C0} between the first and second stage is the Miller-compensation capacitor. It is used to separate the two adjacent poles of OTA, pushing the dominant pole to the origin and making the second pole leave the unit gain bandwidth. In addition, the second pole is further offset by resistor R_{C0}, which makes OTA have the characteristic of monopole and has a better phase characteristic.

In order to optimize the noise and offset voltage, we must ensure that the input transistors of the first stage have a large W/L and area. The large transconductance can effectively lower its flick noise.

3.2 Low-dropout linear regulator

Whether it is portable consumer electronics or large household electrical appliances, in operation the constant changes in load and the other various reasons make the power supply fluctuate in a large range, which is very harmful to the circuit. Especially for high-precision measurement, conversion and detection equipment, the power supply voltage is often required to be stable and has low noise. In order to meet the requirements, almost all electronic devices are powered by a Low-dropout linear regulator (LDO). The LDO has the advantages of simple structure, low cost, low noise and so on. It has been widely used in portable electronic equipment.

3.2.1 Basis of LDO

As a basic power supply module, LDO plays a very important role in analog integrated circuits. The changes in output load and the fluctuation of power supply voltage itself have a great influence on the performance of integrated circuit system. As a result, LDO is used as a linear regulator and is often used in systems with high performance requirements.

LDO adjusts the output voltage by principle of negative feedback, and obtains the stable DC output voltage on the basis of providing a certain output current capability. In normal working state, the output voltage is independent of load, input voltage change and temperature. The minimum input voltage of LDO is determined by the minimum voltage drop of the adjusting transistor, usually 150-300mV.

The basic structure of LDO is shown in Fig. 3.12. LDO is mainly composed of the following parts: bandgap, error amplifier, feedback / phase compensation network, and adjusting transistor. The error amplifier, feedback resistor network, adjusting transistor and phase compensation network constitute the stable output voltage V_{out} of feedback loop.

In LDO, a bandgap provides a voltage reference independent of temperature and power. The error amplifier amplifies the difference between voltage reference and feedback voltage, so that the feedback voltage is basically equal to the voltage reference. Phase compensation network is used to compensate the phase of whole feedback network to ensure the feedback network is stable. The adjusting transistor outputs a stable voltage under the control of error amplifier output. The adjusting transistor is a PMOS transistor, and can also be a NMOS ora NPN transistor. In CMOS process, the PMOS is usually selected.

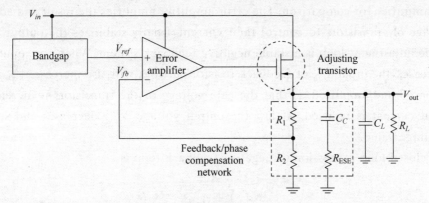

Fig. 3.12   LDO structure

3.2.2 Operational principle

Fig. 3.13 is a LDO curve of the relationship between output voltage and input voltage. The abscissa are the input voltage 0-3.3V, and the ordinate is output voltage. When the input voltage is less than a certain value(1.2-1.8), the output voltage is zero. When the input voltage Vin is greater than it (1.2-1.8V), the output voltage V_{out} increases with V_{in}. When Vin is greater than 2.1V, LDO is in normal operating state and the output voltage is stable at 1.8V.

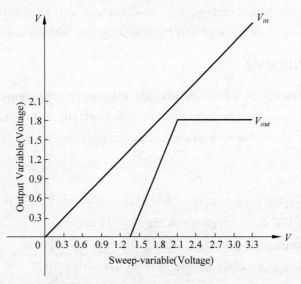

Fig. 3.13   the relationship between output voltage and input voltage

The operational principle of LDO shown in Fig. 3.12 is: when LDO is on power, the start-up circuit in bandgap begins to work to ensure that the whole system begins to operate normally. The bandgap outputs a stable voltage reference V_{ref} which is independent of supply voltage and temperature, while the $R1$ and $R2$ of the feedback / phase compensation network generate feedback voltage V_{fb}. The two voltages are input to error amplifier for comparison. The error amplifier amplifies the result and adjusts the gate voltage of transistors to control their current, finally stabilizes the output voltage. The whole adjustment loop is a stable negative feedback system. When the input voltage V_{out} increases, the input V_{fb} of feedback resistor network will also increase. V_{fb} and V_{ref} are compared and amplified, so that the gate voltage of the transistors is increased, and the output current is reduced. At last the output voltage V_{out} decreases and stays at a stable voltage value.

The closed loop expression of negative feedback loop is

$$V_{out} = \frac{A_{ol}}{1 + A_{ol}\beta} V_{ref} \tag{3-34}$$

where A_{ol} is the open-loop gain of negative feedback loop, and β is the feedback coefficient, and its expression is

$$\beta = \frac{R_2}{R_1 + R_2} \tag{3-35}$$

when $A_{ol}\beta \gg 1$, equation (3-34) can be expressed as

$$V_{out} \approx \frac{V_{ref}}{\beta} = \frac{R_1 + R_2}{R_2} V_{ref} \tag{3-36}$$

It can be seen from (3-35) and (3-36), the output voltage V_{out} of LDO is only related to the voltage reference V_{ref} and the resistor ratio of feedback resistor network. It has nothing to do with the input voltage V_{in}, load current and temperature. Therefore, the required output voltage can be obtained by adjusting the resistor ratio.

3.2.3 Parameter

The main parameter of LDO are divided into static and dynamic parameter. The static parameter includes dropout voltage、quiescent current and efficiency. And dynamic parameter consists of transient response、line regulation、load regulation and power supply rejection ratio.

1. Dropout Voltage

When input voltage V_{in} is less than a certain value of V_{cutoff}, the LDO output zero voltage, and it is in the cutoff region; While V_{in} is between V_{cutoff} and critical voltage V_{Break}, the output voltage is not fixed, and LDO has no adjustment ability; And when V_{in} is greater than V_{Break}, LDO enters normal operating state and the output voltage remains unchanged. We define that the difference between V_{Break} and output voltage is the dropout voltage of LDO, and the dropout voltage is usually 150mV-300mV. Its expression is

$$V_{drop} = V_{in} - V_{Break} \tag{3-37}$$

In general, the smaller dropout voltage is beneficial to improve the conversion efficiency of LDO. However, a too small dropout voltage may cause a poor phase margin and power supply rejection ratio to the whole feedback loop, so a compromise should be considered in LDO design.

2. Quiescent Current

Quiescent current (I_Q) usually refers to the current consumed by the whole LDO when it is not connected to any load, or it is defined as the difference between input current and output current with output load. The expression is shown

$$I_Q = I_{in} - I_{out} \tag{3-38}$$

The quiescent current is mainly composed of the bias current of active devices such as bandgap, error amplifier, active compensation network (if exist), and the current consumed by the feedback resistor network. The smaller quiescent current is beneficial to improve the conversion efficiency of LDO and the life of the battery. In CMOS integrated circuit, the quiescent current of the LDO is in the dozens of μA or smaller.

3. Efficiency

The efficiency of LDO is defined as the percentage of ratio of output power P_{out} to input power P_{in}

$$\beta = \frac{P_{out}}{P_{in}} \cdot 100\% = \frac{I_{out} \cdot V_{out}}{I_{in} \cdot V_{dd}} \cdot 100\% = \frac{(I_{in} - I_Q) \cdot (V_{dd} - V_{drop})}{I_{in} \cdot V_{dd}} \cdot 100\% \tag{3-39}$$

where I_{in} is input current, V_{dd} is supply voltage, I_Q is quiescent current and V_{drop} is dropout voltage.

We can know that from (3-39), the efficiency of LDO related to quiescent current and dropout voltage. The way to improve efficiency is to reduce I_Q and V_{drop}.

4. Transient Response

The transient response of LDO includes two aspects: ① the linear transient response due to supply voltage change; ② the load transient response when load and output voltage suddenly changes. The linear transient response is more important in frequent power up and power down application, while the load transient response is more important when the load current changes frequently. In specific application, the latter occurs in real time, so the load transient response is paid more attention in design.

Fig. 3.14 is the transient response of LDO. The upper half part is the step change of output load current, and the lower part is the output voltage change with load current. The changes in load transient response can be expressed

$$\Delta V_{TR,max} = \frac{I_0 \cdot \Delta t}{C_0} + \Delta V_{ESR} \tag{3-40}$$

I_0 is load current. C_0 is output capacitor. ΔV_{ESR} is the output voltage variation caused by the equivalent series resistor (ESR) of output capacitor. And Δt is step response time. In practical application, the smaller transient response time, the better of LDO performance. By equation (3-40), with the fixed load current increasing the output

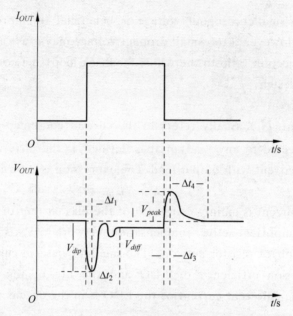

Fig. 3.14　LDO load transient response

capacitor, the closed-loop bandwidth and ESR will reduce the amplitude of load transient response, thereby reducing the load response time. In LDO design, the frequency response method should be adopted to determine the stability of the whole feedback loop first. Because LDO is a closed-loop system with multiple poles and negative feedback, if the design is improper, there may be a stability problem, which will naturally affect its time-domain transient characteristics.

5. Line Regulation

The line regulation of LDO is defined as the variation ratio of output to input voltage when input voltage changes under constant load. It can reflect the ability of LDO to restrain the change of input voltage. Its expression is

$$S_{LR} = \frac{\Delta V_{out}}{\Delta V_{in}} \Big|_{I_{out} = constant} \tag{3-41}$$

In Figure 3.15, it is assumed that the load current is kept at 50mA, and when the input voltage is 2.1V, the LDO enters adjustment state and outputs a 1.8V voltage. The

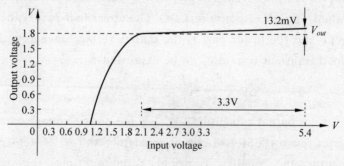

Fig. 3.15　Line regulation of LDO

input voltage is gradually increased to 5.4V, and the output voltage varies from 1.8V to 1.8132V, so the line regulation of LDO is

$$S_{LR} = \frac{0.0132}{(5.4-2.1)}\bigg|_{I_{out}=50\text{mA}} = 4\text{mV/V}$$

6. Load Regulation

The load regulation of LDO is defined as the ratio of the output voltage variation ΔV_{out} to the load current variation ΔI_{load} when input voltage V_{in} remains unchanged. The load regulation reflects the influence of load current on output voltage, and the smaller, the better it is. Its expression is:

$$S_{LOR} = \frac{\Delta V_{out}}{\Delta I_{load}}\bigg|_{V_{in}=constant} \quad (3\text{-}42)$$

Figure 3.16 gives a schematic of load regulation, where I_{load} and ΔI_{load} are load current and variation value respectively, and V_{out} and ΔV_{out} are output voltage and output voltage variation value respectively.

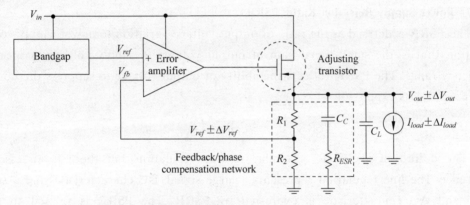

Fig. 3.16 Circuit diagram for calculating load regulation

When the output load current changes to $I_{load} + \Delta I_{load}$, the output voltage changes to $V_{out} + \Delta V_{out}$, which makes the output voltage variation of feedback resistor network is delta ΔV_{fb}

$$\Delta V_{fb} = \Delta V_{out} \cdot \frac{R2}{R1+R2} \quad (3\text{-}43)$$

and ΔI_{load} is

$$\Delta I_{load} = g_{mEA} \cdot g_{mp} \cdot \Delta V_{fb} \quad (3\text{-}44)$$

g_{mEA} and g_{mp} are the transconductance of error amplifier and adjusting transistor. Putting (3-43) and (3-44) into (3-42), the expression of load regulation is

$$S_{LOR} = \frac{1}{g_{mEA}g_{mp}} \cdot \frac{R_1+R_2}{R_2}\bigg|_{V_{in}=constant} \quad (3\text{-}45)$$

From equation (3-45), we can see that we can improve load regulation by increasing the transconductance of error amplifier and adjusting transistor, and the feedback coefficient of feedback resistor network.

In Fig. 3.17, assuming the input voltage $V_{in}=3.3\text{V}$, the load current is changed from 0 to 50mA, and the output voltage varies about 1.32mV near 1.8V, then load regulation is

$$S_{LOR} = \frac{1.32\text{mV}}{50\text{mA}}\bigg|_{V_{in}=3.3\text{V}} = 0.0264\text{V/A}\bigg|_{V_{in}=3.3\text{V}} \tag{3-46}$$

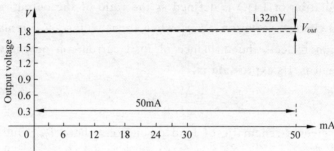

Fig. 3.17 Load regulation

7. Power Supply Rejection Ratio (PSRR)

The PSRR is defined as the ratio of output voltage variation to power supply voltage variation, that is, the small-signal gain of output voltage to supply voltage in a certain frequency range. The PSRR reflects the ability of output voltage to suppress the noise of power supply. Its expression is

$$\text{PSRR}\big|_{\text{Hz}} = 20 \cdot \log_{10}\left(\frac{\Delta V_{out}}{\Delta V_{dd}}\right) \text{dB} \tag{3-47}$$

The definition of PSRR is similar to line regulation, but there is an essential difference. The line regulation represents a large signal, DC characteristic, and a small-signal and AC characteristic is expressed by PSRR. The PSRR is related to LDO structure, PSRR of bandgap and error amplifier. Fig. 3.18 is a typical PSRR curve. Because LDO mainly provides DC voltage, the PSRR within 10KHz is more important. Usually high-precision circuits are required for PSRR within 100kHz.

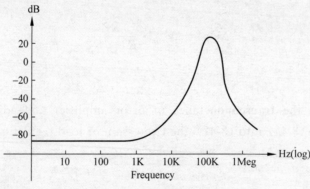

Fig. 3.18 PSRR curve

3.2.4 Stability analysis

LDO is a closed-loop system with negative feedback. The system outputs feedback voltage to input terminal and compares with the voltage reference for adjusting, finally making the output stable on a expected value. Due to the complexity of feedback system and existence of multiple zero-poles, there is a stability problem in itself. If the system is not designed properly, it will cause the whole negative feedback system to oscillate. Therefore we must analyze the open-loop amplitude-frequency and phase-frequency characteristics, so as to ensure the stability.

Usually, the condition of ensuring stability is that the phase margin of open-loop characteristic is greater than 45 degrees. Considering the transient characteristics, overshoot and other properties, the 60 degree is an ideal value.

General LDO compensates the poles to achieve a better open-loop phase margin by using the equivalent series resistor(ESR) of off-chip capacitors, as Fig. 3.19 shows.

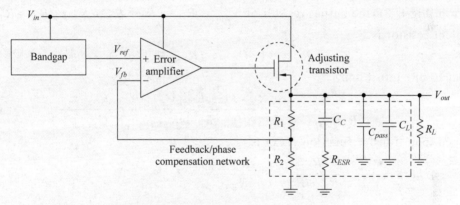

Fig. 3.19 LDO with off-chip capacitors

In Fig. 3.20, $R1$ and $R2$ are feedback network resistor, C_C is off-chip compensation capacitor, and R_{ESR} is the equivalent series resistor of compensation capacitor. The range of R_{ESR} is $10m\Omega$-1Ω. C_C generally takes a value of $0.1\mu F$-$10\mu F$, and a tantalum capacitor with good performance is generally used. Because the value of C_C is large, the output node Vout is the dominant pole of LDO. C_{pass} is a bypass capacitor. Usually, when the load current changes suddenly and the output voltage generates large ripple, C_{pass} will reduce the ripple amplitude and make the output voltage stable quickly. The load capacitor CL is usually smaller than C_C and C_{pass}. For simplicity, it is not considered in the following analysis.

When analyzing the small-signal characteristic of off-chip compensation LDO, we need to break a certain point in the feedback loop. Start from this point, analyzing the components of system sequentially, and finally returning to the starting point, as Fig. 3.20 shows.

In Fig. 3.21, we disconnect the feedback network and define error amplifier input

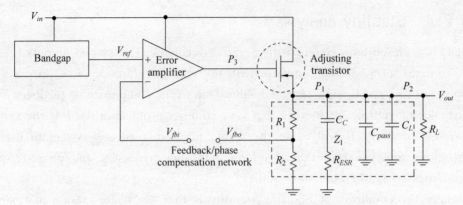

Fig. 3.20 Analysis of small-signal characteristic for off-chip compensation LDO

V_{fbi} as the starting point of analysis, and the feedback resistor output is V_{fbo}. So the equivalent output resistor of error amplifier is R_{EA}, and the gate equivalent input capacitor of adjusting transistor is C_{GP}, and the source-drain equivalent resistor is R_{DSP}.

From Fig.3.21, the output resistor of V_{out} is $R_o = (R_1 + R_2) // R_{DSP} // R_L \approx R_{DSP}$, its equivalent resistor is Z_{out},

$$Z_{out} = R_o // (1/sC_b) // (R_{ESR} + 1/sC_c) \qquad (3\text{-}48)$$

put (3-48) into (3-49)

$$Z_{out} = \frac{R_o \cdot (1 + sR_{ESR}C_c)}{s^2 R_o R_{ESR} C_c C_{pass} + s[C_c(R_o + R_{ESR}) + R_o C_{pass}] + 1} \qquad (3\text{-}49)$$

The open-loop transfer function $H(s)$ is

$$H(s) = \frac{V_{fbo}}{V_{fbi}} = \frac{R_2}{R_1 + R_2} \cdot \frac{g_{EA} R_{EA} g_{MP}}{1 + sR_{EA}C_{GP}} \cdot Z_{out}$$

$$H(s) = \frac{R_2}{R_1 + R_2} \cdot$$

$$\frac{g_{EA} R_{EA} g_{MP}(1 + sR_{ESR}C_c)}{[1 + s(R_0 + R_{ESR})C_c] \cdot [1 + s(R_0 // R_{ESR})C_{pass}] \cdot (1 + sR_{EA}C_{GP})} \qquad (3\text{-}50)$$

Considering $R_0 \approx R_{DSP} \ll R1 + R2, R_0 \gg R_{ESR}$, the poles and zeroes of entire LDO open-loop system are

$$P_1 = -\frac{1}{(R_0 + R_{ESR}) \cdot C_c} = -\frac{1}{R_{DSP} \cdot C_c} \qquad (3\text{-}51)$$

$$P_2 = -\frac{1}{(R_0 // R_{ESR}) \cdot C_{pass}} = -\frac{1}{R_{ESR} \cdot C_{pass}} \qquad (3\text{-}52)$$

$$P_3 = -\frac{1}{R_{EA} \cdot C_{GP}} \qquad (3\text{-}53)$$

$$Z_1 = -\frac{1}{R_{ESR} \cdot C_c} \qquad (3\text{-}54)$$

In general, $P1 < P2 < P3$, and $P1 < Z1 < P2$. The frequency response is shown in Fig.3.47.

The system's open-loop phase margin formula shows that if the phase margin is at

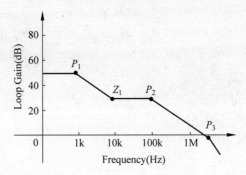

Fig. 3.21 Frequency response

least 60 degrees, there is only one pole in the unit gain bandwidth (UGB). Therefore it shows the characteristics of single pole system, and the non dominant pole is outside 2.2 times of UGB. In Fig. 3.21, if the phase margin is up to 60 degrees, Z1 must be adjacent to P2 and P3>2.2UGB, which means that zero Z1 must be used to compensate for the non dominant pole P2. Furthermore the choice of the equivalent series resistance R_{ESR} must also be in a reasonable range. If R_{ESR} is too large, the zero Z1 near the low frequency will cause UGB to increase. It may cause pole P3 to go into UGB, and the phase margin becomes smaller, which results in the instability of system; But If the R_{ESR} is too small, the zero Z1 moves to high frequency, which may move out of UGB and can not achieve compensation. So R_{ESR} has an important impact on system's stability. In fact, due to the large size of adjusting transistor, the gate-source capacitor also constitutes a zero.

Because of error amplifier with large output impedance (R_{EA}) and gate capacitor of adjusting transistor (C_{GP}), its output has a large time constant. Since its large and small-signal response time are subject to these restrictions, LDO as shown in Fig. 3.19 is not commonly used. In order to improve the accuracy and transient characteristics of LDO, a buffer is usually inserted between error amplifier and adjusting transistor.

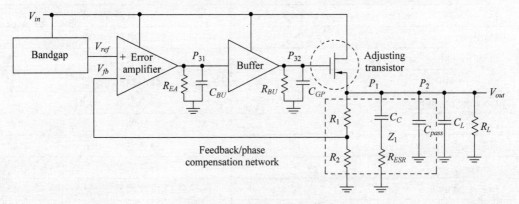

Fig. 3.22 LDO with buffer

As shown in Fig. 3.22, the voltage buffer isolates error amplifier from adjusting transistor, and divides the original pole P3 into two poles: P31 and P32.

$$P_{31} = \frac{1}{R_{EA}C_{BU}} \tag{3-55}$$

$$P_{32} = \frac{1}{R_{BU}C_{GP}} \tag{3-56}$$

Figure 3.23 is the amplitude-frequency characteristic of LDO with buffer. When the buffer is added, the poles $P31$ and $P32$ are greater than the original pole $P3$, so the system is more stable and improves its transient response time greatly.

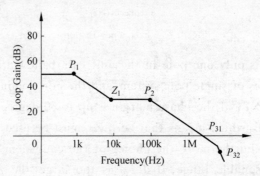

Fig. 3.23 amplitude-frequency characteristic of LDO with buffer

3.3 Technical words and phrases

3.3.1 Terminology

bandgap	带隙基准源
low-dropout linear regulator(LDO)	低压差线性稳压器
positive (negetive) temperature coefficient	正(负)温度系数
subthreshold region	亚阈值区
Boltzmann constant	玻尔兹曼常数
proportional to absolute temperature(PTAT)	与绝对温度成正比
start-up	启动电路
error amplifier	误差放大器
static parameter	静态参数
dynamic parameter	动态参数
dropout voltage	电压差
quiescent current	静态电流
efficiency	转换效率
line regulation	线性调整率
load regulation	负载调整率
equivalent series resistor(ESR)	等效串联电阻

3.3.2 Note to the text

(1) The bandgap is fully compatible with standard CMOS process and can work at low power supply voltage. In addition, it has low temperature drift, low noise and high power supply rejection ratio, which can meet the requirements of most electronic systems.

带隙基准源与标准 CMOS 工艺完全兼容,可以工作于低电源电压下等优点,另外还具有低温度漂移、低噪声和较高的电源抑制比等性能,能够满足大部分电子系统的要求。

(2) In semiconductor technology, bipolar transistors can provide physical quantities of positive and negative temperature coefficient respectively.

在半导体工艺中,双极晶体管能够分别提供正、负温度系数的物理量。

(3) If two identical bipolar transistors are biased at different collector current, the difference between their base-emitter voltage is proportional to the absolute temperature.

如果两个相同的双极晶体管在不同的集电极电流偏置情况下,那么它们的基极-发射极电压的差值与绝对温度成正比。

(4) There are two main kinds of circuit structure to complete adding, one is to add the two through an OPA. And its output is the voltage reference. The other is to generate a current proportional to absolute temperature(PTAT), which can be converted into a voltage through a resistor. This voltage naturally has a positive temperature coefficient, which is then added to the base-emitter voltage V_{BE}.

完成这种相加的电路结构目前主要有两种,一种是通过运算放大器将两者进行相加,输出即为基准电压;另一种是先产生与温度成正比(PTAT)的电流,通过电阻转换成电压,这个电压自然具有正温度系数,然后与二极管的基极-发射极电压 V_{BE} 相加获得。

(5) In order to solve the start-up problem, an additional circuit is needed. The basic requirement of the start-up circuit is that after the power supply is stable, when circuit is in the "zero current" working state, the start-up circuit gives an stimulus to the internal circuit node, which forces it to get rid of the "zero current" working state and fall into the normal working mode.

为了解决电路的启动问题,需要加入额外电路,使得存在启动问题的电路摆脱"零电流"工作状态进入正常工作模式,对启动电路的基本要求是电源电压稳定后,待启动电路处于"零电流"工作状态时,启动电路给内部电路某一节点激励信号,迫使待启动电路摆脱"零电流"工作状态,而进入正常工作模式。

(6) In general, the smaller dropout voltage is beneficial to improve the conversion

efficiency of LDO. However, a too small dropout voltage may cause a poor phase margin and power supply rejection ratio to the whole feedback loop, so a compromise should be considered in LDO design.

通常情况下,要求 LDO 的电压差越小越好,以提高整体电路的转换效率;但是过小电压差可能会造成整个反馈环路的相位裕度、电源抑制比很差,所以在系统设计时要折中考虑。

(7) The line regulation of LDO is defined as the variation ratio of output to input voltage when input voltage changes under constant load.

LDO 的线性调整率定义为在负载保持恒定的情况下,输入电压发生变化时,输出电压变化量与输入电压变化量的比值。

(8) The definition of PSRR is similar to line regulation, but there is an essential difference. The line regulation represents a large signal, DC characteristic, and a small-signal and AC characteristic is expressed by PSRR.

电源抑制比的定义与线性调整率虽然相似,但是有本质的区别,线性调整率表示的是大信号、直流特性,而电源抑制比表示的小信号、交流特性。

(9) Usually, the condition of ensuring stability is that the phase margin of open-loop characteristic is greater than 45 degrees. Considering the transient characteristics, overshoot and other properties, the 60 degree is an ideal value.

通常情况下,确保负反馈系统稳定的条件是其开环特性的相位裕度大于 45°,考虑到瞬态建立特性、过冲以及其他性能,相位裕度达到 60°是一个比较理想的值。

(10) In order to improve the accuracy and transient characteristics of LDO, a buffer is usually inserted between error amplifier and adjusting transistor.

为了提高 LDO 的精度和瞬态特性,一般在误差放大器和调整晶体管之间插入一个缓冲器。

Chapter 4 Analog filter

Filter is a circuit that adjusts the frequency characteristic of signals. It amplifies or attenuates the signals of different frequency intervals, so as to realize the signals selection. Analog filter circuit exists in all analog systems. In general, a system needs many different filters to eliminate noise signals, suppress interference signals, extract useful signals and adjust signal phase.

4.1 The classification of analog filters

With different classification methods, analog filters can be divided into different types. Base on the selection of active devices, it can be divided into passive filters and active filters. According to different functions, it includes low-pass filter, high-pass filter, band-pass filter, band-stop filter and all-pass filter. In the light of approximation methods, it consists of Butterworth, Chebyshev, Bessel and elliptic filter.

4.1.1 According to the components

The analog filters are divided into passive filters and active filters based on the differences in the use of electronic components. Passive filters are implemented by passive devices such as resistors, capacitors and inductors. It is the prototype of active filter. Its main feature is that the noise coefficient is large and the signal has a certain attenuation during transmission process. Fig. 4.1 demonstrates an implementation of a passive filter. Active filters are composed of active devices (OTA) and passive devices (resistor and capacitor). Due to its larger area, smaller inductance value and lower quality factor of passive inductor in integrated circuit, in general, it is not used in active filter. But the frequency characteristic of inductor can be simulated by OTA and capacitor. Active filter has the advantages of low noise and small area. A active filter is shown in Fig. 4.2.

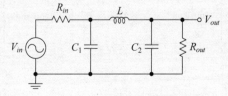

Fig. 4.1 Passive filters

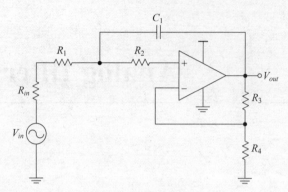

Fig. 4.2 Active filters

4.1.2 According to the functions

Low-pass filter: f_p is defined as low-pass cut-off frequency. Signal below this frequency remains unchanged, and above this frequency is attenuated. Ideally, the signal amplitude can be attenuated to zero. Its amplitude-frequency characteristic is shown in Fig. 4.3(a).

High-pass filter: f_p is the high-pass cut-off frequency. Signal below this frequency is attenuated to zero, and above this frequency remains unchanged. Its amplitude-frequency characteristic is shown in Fig. 4.3(b).

Band-pass filter: the low and high cut-off frequencies f_{p1}, f_{p2} are defined respectively, and $f_{p1} < f_{p2}$. The signal whose frequency is lower than f_{p1} and higher than f_{p2}, can be reduced to zero in ideal case, and the signal between f_{p1} and f_{p2} remains the same. The amplitude-frequency characteristic can be seen in Fig. 4.3(c).

Band-stop filter: the low and high cut-off frequencies is also f_{p1}, f_{p2} respectively, and $f_{p1} < f_{p2}$. The signal whose frequency is lower than f_{p1} and higher than f_{p2}, remains the same, and the signal between f_{p1} and f_{p2} can be reduced to zero in ideal case. The amplitude-frequency characteristic can be seen in Fig. 4.3(d).

All-pass filter: The amplitude-frequency characteristic of all-pass filter is kept constant as 1 in all frequency range, that is, all signals are not attenuated. The phase-frequency characteristic remains linear throughout the frequency range. Its amplitude-frequency and phase-frequency characteristic are shown in Fig. 4.3(e).

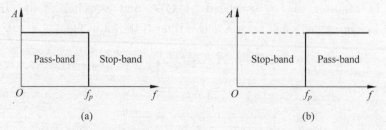

Fig. 4.3 Amplitude-frequency characteristics of analog filters

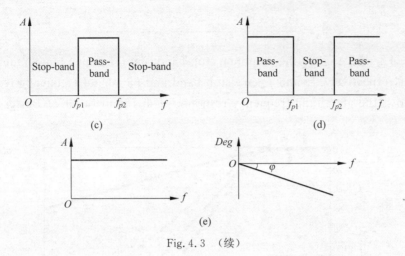

Fig. 4.3 （续）

4.1.3 According to the approximation functions

A physical realization of an ideal filter transfer function is impossible. Hence, a frequency scheme is given, which is dependent on many system parameters and trade-offs. Within this frequency scheme the real filter transfer function has to be located. Hence various approximations for the ideal filter transfer functions are realizable. Important and popular approximation functions are Butterworth approximations, Chebyshev approximations, elliptic approximations, and Bessel approximations.

1. Butterworth approximation

The Butterworth filter approximation is demonstrated on the normalized low-pass Butterworth filter. High-pass, band-pass, and band-stop filters can be realized by using the appropriate transformation. The normalized Butterworth function of Nth-order is given by (4-1)

$$|H(j\omega)| = \frac{1}{\sqrt{1+\varepsilon^2(\omega/\omega_c)^{2n}}} \qquad (4-1)$$

ω_c is the cut-off frequency. The magnitude function of a Butterworth low-pass filter is monotonically decreasing for $\omega > 0$ and the maximum of $H(j\omega)$ is at $\omega = 0$. A Nth-order Butterworth low-pass filter has a maximally flat magnitude function. It also shows an overshoot in the step response in the time domain, which worsens at increasing filter-order N.

The distribution of the poles and zeros in the s-plane are characteristic for filters and can also be used for filter identification. A normalized Butterworth low-pass filter has its poles in the left half plane on the circumference of the unity circle with the center in the point of origin of the s-plane. The poles are spread equidistant over the half circle in the left half s-plane.

The order L can be represented by

$$L = \frac{\log_{10}((10^{0.1A_{SB}} - 1)/(10^{0.1A_{PB}} - 1))}{2\log_{10}(f_s/f_p)} \tag{4-2}$$

where A_{SB} and A_{PB} are the minimum stop-band suppression and maximum pass-band attenuation, respectively. f_s and f_p are stop-band and pass-band frequency respectively. Fig. 4.4 shows the amplitude-frequency response of different order of Butterworth low-pass filter.

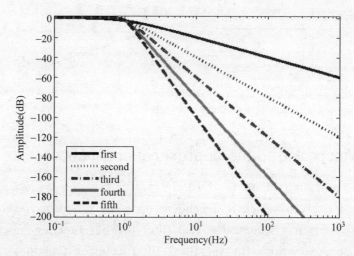

Fig. 4.4 The amplitude-frequency response of different order of Butterworth low-pass filter

2. Chebyshev approximation

Chebyshev filters offer a frequency response that approximates the ideal low-pass filter more precisely than a Butterworth filter. The transfer function has only poles and lack of any finite zeros. The pass-band of a Chebyshev filter exhibits a passband ripple, which ranges between two constant values. The ripple amplitude can be adjusted freely. The ripple amplitude is direct proportional to the filter slope in the transition-band (filter selectivity) and the overshoot of the step response in the time domain. The greater the pass-band ripple, the higher the filter selectivity and the overshoot. The number of ripples depends on the order N of the filter. For frequencies greater than the cut-off frequency the filter has a monotonically decreasing magnitude function similar to Butterworth filters.

If a L-order Chebyshev filter with cut-off frequency ω_c, and it satisfies equation (4-3), the amplitude-frequency response is (4-4)

$$A_{max} = 10\log_{10}(1 + \varepsilon^2) \tag{4-3}$$

$$|H(j\omega)| = \frac{1}{\sqrt{1 + \varepsilon^2 C_L^2(\omega/\omega_c)}} \tag{4-4}$$

in (4-4), $C_L(\omega/\omega_c)$ is called the L-order Chebyshev polynomial, which is defined as

$$C_L(\omega/\omega_c \leqslant 1) = \cos[L\arccos(\omega/\omega_c)]$$

$$C_L(\omega/\omega_c \geqslant 1) = \cosh[L\arccos(\omega/\omega_c)] \tag{4-5}$$

And $C_n^2(\omega/\omega_c \leqslant 1) \leqslant 1, C_n^2(\omega/\omega_c \geqslant 1) \geqslant 1$. The filter order required by Chebyshev approximation is shown in (4-6).

$$L = \frac{\cosh^{-1}\sqrt{(10^{0.1A_{SB}}-1)/(10^{0.1A_{PB}}-1)}}{\cosh^{-1}(f_s/f_p)} \tag{4-6}$$

The 1dB-ripple Chebyshev approximation response for different orders is shown in Fig. 4.5.

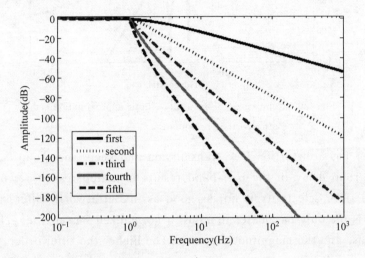

Fig. 4.5 The 1dB-ripple Chebyshev approximation response for different orders

There is a inverse Chebyshev approximation which has complementary properties of the Chebyshev filter concerning the magnitude response. In the pass-band the magnitude function is monotonically decreasing for $\omega > 0$ and the equiripple appears in the stop-band. The filter selectivity is not so good as in Chebyshev filters. The zeros in the filter transfer function bring an additional realization effort.

3. Elliptic approximation

Elliptic filters have an equal ripple in the pass-and in the stop-band. Therefore elliptic filter is also called double Chebyshev filter. The two ripples are individually adjustable. The transition between pass-band and stop-band show a high filter selectivity. The sharp filter edge is realized by a transfer function using a balanced combination of poles and zeros. The magnitude function of an elliptic filter is the best approximation of the ideal low-pass filter compared to Butterworth and Chebyshev filters. For the same roll-off in transition band, the required order of elliptical filter is the least. Fig. 4.6 is the comparison of 3rd-order elliptic approximation and Butterworth approximation.

4. Bessel approximation

In contrast to Butterworth or Chebyshev filters, which approximate the magnitude function of an ideal low-pass filter, the Bessel filter approximates the phase response. Hence, the Bessel filter is also called maximally flat group delay filter. In the pass-band

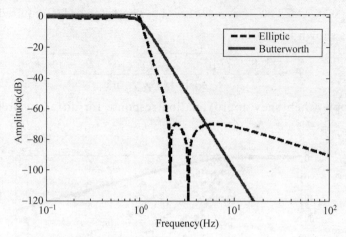

Fig. 4.6 Comparison of 3rd-order elliptic approximation and Butterworth approximation

the Bessel filter has a distortion free transmission and keeps the group delay constant. The constant group delay in the pass-band results in a step response, which shows no overshoot. The filter selectivity is not as good as in Butterworth filter structures. The filter order N is the only parameter to adjust a normalized Bessel filter. The value of N defines the phase and the magnitude response. The higher the filter order, the larger the frequency range with constant group delay and the higher the filter selectivity.

Figures 4.7 and 4.8 are the frequency characteristics and step response curves of 5th-order filters with the same cut-off frequency and different approximations. It can be seen that in four different approximations Bessel approximation has the best phase characteristics with the smallest overshoot. And the Elliptic approximation has the sharpest transition, while Butterworth approximation has the maximum flatness in the passband.

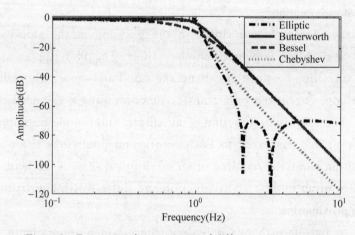

Fig. 4.7 Frequency characteristics of different approximations

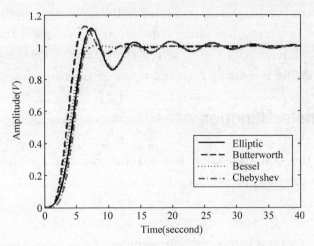

Fig. 4.8 Step response of different approximations

4.2 Amplitude-frequency characteristics

The frequency characteristic of ideal filter is impossible. The real filter needs to be realized by approximation method, and its amplitude-frequency characteristic is shown in Fig. 4.9.

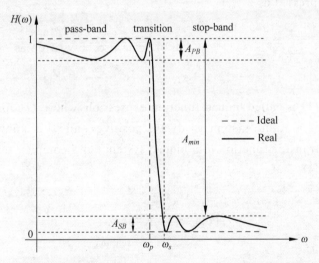

Fig. 4.9 Amplitude-frequency characteristic of analog filter

It can be seen that the passband of filter is a frequency range with little attenuation or less attenuation of signal. The passband gain does not need to be a fixed value, and it can be a maximum range of amplitude, which is called the maximum ripple of A_{PB}. The passband frequency of filter shown in Fig. 4.9 extends from the DC to ω_p, and ω_p is called cut-off frequency. The amplitude characteristic curve will fall into the stopband quickly after passing through. ω_p. The stop-band is frequency region with amplitude attenuation that starts from ω_s. A_{min} is defined as the minimum amplitude attenuation or stopband rejection. The frequency range from ω_p to ω_s is called transition. The ratio ω_s/ω_p represents the selectivity factor or roll-off. Generally speaking, we can increase the transition roll-off by increasing passband ripple or the filter order.

4.3 Transfer function

The transfer function of analog filter is the rational function of the complex frequency s, which can be expressed as

$$H(s) = K \frac{(s-z_1)(s-z_2)(s-z_3)\cdots(s-z_n)}{(s-p_1)(s-p_2)(s-p_3)\cdots(s-p_m)} \tag{4-7}$$

where K is a constant, known as filter gain. $z_1, z_2 ..., z_n$ is zero of filter, and p_1, $p_2, ..., p_m$ is the pole. The zero and pole can be a real or complex number. When the zero and pole are complex numbers, they exist in the form of conjugacy, as shown in equation (4-8) and (4-9)

$$s_1 = \text{Re}(s) + j\text{Im}(s) \tag{4-8}$$

$$s_{1,conj} = \text{Re}(s) - j\text{Im}(s) \tag{4-9}$$

The 2nd-order polynomial of transfer function expressed by the conjugate complex number can be expressed as

$$(s-s_1)(s-s_{1,conj}) = s^2 - (s_1 + s_{1,conj})s + s_1 s_{1,conj} = s^2 + as + \omega^2 \tag{4-10}$$

and $a = s_1 + s_{1,conj}$, $\omega^2 = s_1 s_{1,conj}$, (4-10) can also be written as

$$(s-s_1)(s-s_{1,conj}) = s^2 + \frac{\omega}{Q}s + \omega^2 \tag{4-11}$$

Equation (4-11) is called biquad function expression, where Q represents the quality factor. Q value of a stable system is always positive, and ω is known as the angular frequency. If the transfer function of a biquad system has a complex zero and pole, it can be expressed as

$$H(s) = \frac{s^2 + \frac{\omega_z}{Q_z}s + \omega_z^2}{s^2 + \frac{\omega_p}{Q_p}s + \omega_p^2} \tag{4-12}$$

ω_z and ω_p are zero and pole of biquad system. Q_z and Q_p are quality factor of zero and pole. Equation (4-12) can also be expressed as

$$H(s) = \frac{As^2 + Bs + C\omega_p^2}{s^2 + \frac{\omega_p}{Q_p}s + \omega_p^2} \tag{4-13}$$

The constant coefficients A, B and C are real numbers, and their defined values can determine the frequency characteristics of low-pass, high-pass, band-pass, band-stop and all-pass, respectively, for the 2nd order transfer function.

4.3.1 Low-pass transfer function

The biquad low-pass transfer function can be expressed

$$H_{LP}(s) = \frac{\omega_p^2}{s^2 + \dfrac{\omega_p}{Q_p}s + \omega_p^2} \tag{4-14}$$

ω_p represents the cut-off frequency of low-pass transfer function. Q_p represents the quality factor. At low frequency, the quadratic and linear term of denominator are negligible, and the amplitude-frequency characteristic can be considered approximately 1. When $\omega \gg \omega_p$, the amplitude-frequency characteristic decreases with the angular frequency ω^2, and decreases at the rate of -40dB/dec on bode plot to achieve the low-pass characteristic. Its amplitude-frequency and phase-frequency characteristic are shown in Fig. 4.10.

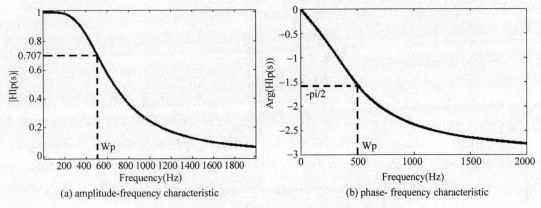

(a) amplitude-frequency characteristic (b) phase-frequency characteristic

Fig. 4.10 The biquad low-pass transfer function

4.3.2 High-pass transfer function

The biquad high-pass transfer function can be expressed

$$H_{HP}(s) = \frac{s^2}{s^2 + \dfrac{\omega_p}{Q_p}s + \omega_p^2} \tag{4-15}$$

ω_p represents the cut-off frequency of high-pass transfer function. Q_p represents the quality factor. At low frequency, the quadratic and linear term of denominator are also negligible, and the amplitude-frequency characteristic can be considered approximately 0. When $\omega \gg \omega_p$, the amplitude-frequency characteristic approximate 1 to achieve the high-pass characteristic. Its amplitude-frequency and phase-frequency characteristic are shown in Fig. 4.11.

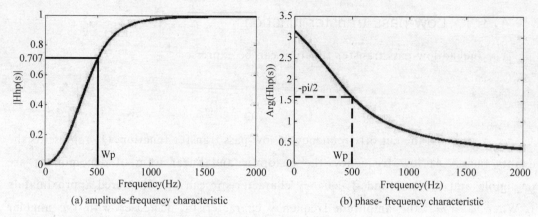

(a) amplitude-frequency characteristic (b) phase- frequency characteristic

Fig. 4.11 The biquad high-pass transfer function

4.3.3 Band-pass transfer function

Its transfer function can be expressed

$$H_{BP}(s) = \frac{\frac{\omega_p}{Q_p}s}{s^2 + \frac{\omega_p}{Q_p}s + \omega_p^2} \tag{4-16}$$

At low frequency $s = 0$, its amplitude-frequency characteristic is 0; At high frequency when $s \to \infty$, its amplitude-frequency characteristic is 0 as well. While $s = j\omega_p$, The amplitude-frequency characteristic is the maximum, and the band-pass characteristic is realized. ω_p is called the resonant frequency. When the amplitude-frequency characteristic is $1/\sqrt{2}$ times of the maximum, the corresponding frequency is the cut-off frequency ω_1 and ω_2. And there are $\omega_{low,3dB} = \min(\omega_1, \omega_2)$ and $\omega_{high,3dB} = \max(\omega_1, \omega_2)$. Its bandwidth is expressed as (4-17) and shown in Fig. 4.12.

$$\omega_{bw} = \omega_{high,3dB} - \omega_{low,3dB} \tag{4-17}$$

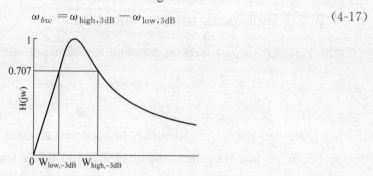

Fig. 4.12 Bandwidth of biquad band-pass filter

Fig. 4.13 shows the amplitude-frequency characteristics of biquad band-pass filters with the same resonant frequency 500Hz and different Q values. It can be seen that the biquad band-pass filter has smaller bandwidth with greater Q value, or it means that the greater Q value, the better the selectivity of the filter.

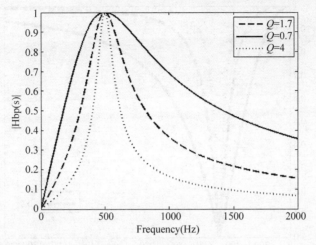

Fig. 4.13 Amplitude-frequency characteristics of biquad band-pass filters with different Q values

4.3.4 Band-stop transfer function

Its transfer function can be expressed

$$H_{BR}(s) = \frac{s^2 + \omega_p^2}{s^2 + \dfrac{\omega_p}{Q_p}s + \omega_p^2} \qquad (4\text{-}18)$$

At low frequency $s = 0$, its amplitude-frequency characteristic is 1; At high frequency when $s \to \infty$, its amplitude-frequency characteristic is 1 as well. While $s = j\omega_p$, The amplitude-frequency characteristic is the minimum, and the band-stop characteristic is realized. When the amplitude-frequency characteristic is $1/\sqrt{2}$ times of the maximum, the corresponding frequency is the cut-off frequency ω_1 and ω_2. And there are $\omega_{\text{low,3dB}} = \min(\omega_1, \omega_2)$ and $\omega_{\text{high,3dB}} = \max(\omega_1, \omega_2)$. Its bandwidth is expressed as (4-19) and shown in Fig. 4.14.

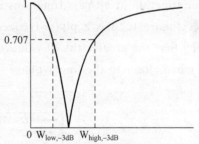

Fig. 4.14 Bandwidth of biquad band-stop filter

$$\omega_{bw} = \omega_{\text{high,3dB}} - \omega_{\text{low,3dB}} \qquad (4\text{-}19)$$

Fig. 4.15 shows the amplitude-frequency characteristics of biquad band-stop filters with the same resonant frequency 500Hz and different Q values. It can be seen that the biquad band-stop filter has smaller bandwidth with greater Q value, or it means that the greater Q value, the better the selectivity of the filter.

4.3.5 All-pass transfer function

Its transfer function can be expressed

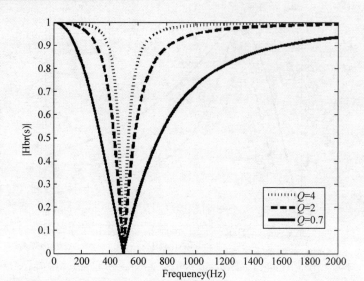

Fig. 4.15 Amplitude-frequency characteristics of biquad band-stop filters with different Q values

$$H_{AP}(s) = \frac{s^2 - \dfrac{\omega_p}{Q_p}s + \omega_p^2}{s^2 + \dfrac{\omega_p}{Q_p}s + \omega_p^2} \tag{4-20}$$

The amplitude-frequency characteristic of all-pass functions is 1 at the whole frequency. For all-pass functions, the phase-frequency characteristic is more concerned. The function of all-pass functions is phase correction.

Figure 4.16 is a phase-frequency characteristic of biquad all-pass filter. It can be seen that the greater the Q value, the better the phase-frequency characteristic is. And it is much closer to the linear phase.

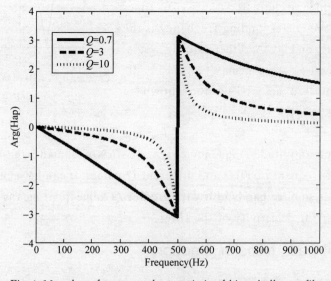

Fig. 4.16 phase-frequency characteristic of biquad all-pass filter

4.4 Implementation of analog filter circuit

All kinds of active filters described above include integrators, active resistors and active inductors. This section is a brief introduction to these modules.

4.4.1 Active RC integrator

The integrator, realized by resistor, capacitor and OTA, is called an active RC integrator. Its structure is shown in Fig. 4.17.

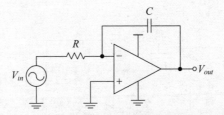

Fig. 4.17　Active RC integrator

Assuming that the OTA is ideal, the Kirchhoff voltage and current law are applied to the negative input, and the transfer function of active RC integrator is

$$\frac{V_{out}}{V_{in}}(s) = \frac{1}{sRC} \qquad (4\text{-}21)$$

If the non-ideal factors such as limited gain and bandwidth of OTA are taken into account, the transfer function of active RC integrator shown in Fig. 4.17 can be approximately expressed as

$$\frac{V_{out}}{V_{in}}(s) = \frac{1}{sRC(1+s/f_t)} = \frac{1}{s\tau(1+s/f_t)} \qquad (4\text{-}22)$$

In equation (4-11), f_t is the UGW of OTA, which can be approximated as the product of its low-frequency gain and the dominant-pole frequency. Compared with (4-21), the denominator in equation (4-22) has a s/f_t term, which can be considered as a loss caused by the limited UGW of OTA. In active RC integrator, the loss increases with signal frequency. And the frequency response will also has a bigger error, which limits its application in broadband.

4.4.2 MOS-C integrator

Because the MOS transistor operating in linear region can be used as a resistor, and the resistance can be continuously adjustable according to gate voltage. So the application of this adjustable resistor in active RC integrator can improve its frequency accuracy. Equation (4-23) is the current-voltage relationship of a MOS transistor working in linear region. The 2nd-order effects of channel length modulation and body effect are not considered.

$$I_{ds} = \frac{1}{2}\mu C_{ox} \frac{W}{L}[2(V_{gs} - V_{th})V_{ds} - V_{ds}^2] \tag{4-23}$$

V_{gs} is the gate-source voltage, and V_{ds} is drain-source voltage, V_{th} is threshold voltage. From the derivative of (4-23) with respect to V_{ds}, the resistance of the transistor can be obtained

$$R_{ds} = \frac{1}{\mu C_{ox} W/L[(V_{gs} - V_{th}) - V_{ds}]} \tag{4-24}$$

It is known that when V_{ds} is very small, R_{ds} of the transistor is approximately proportional to gate-source voltage, and can form a linear resistor. Fig. 4.18 is the structure of MOS-C integrator.

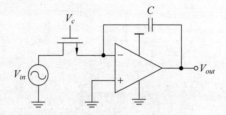

Fig. 4.18 MOS-C integrator

In general, the filter made up of MOS-C integrator can be more accurate than the active RC filter. However, due to the smaller linear interval used for adjustable resistors, the filter's dynamic range is severely limited. Meanwhile, the limited gain and bandwidth of OTA also exist.

4.4.3 Gm-C integrator

When an OTA is used as an active device, its limited gain and bandwidth seriously restrict the operating frequency of filter. Therefore, when frequency is high, the transconductance device is used as an active device. The integrator, consisting of a transconductance and capacitor, is called a Gm-C integrator, and its structure is shown in Fig. 4.19.

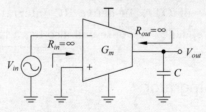

Fig. 4.19 Gm-C integrator

Its transfer function is

$$\frac{V_{out}}{V_{in}}(s) = \frac{G_m}{sC} \tag{4-25}$$

The significant difference between Gm-C integrator and active RC integrator shown is that the Gm-C integrator is open loop. The transconductance Gm output current and does not need low-impedance output stage. Its internal circuit does not need inter-stage capacitor compensation as well. The output pole is the dominant pole, so it has good high-frequency characteristics. Because the Gm-C integrator is an open loop, the biggest problem of this structure is the linear input range of the transconductance. Gm-C integrators usually need to linearize the transconductance. The methods of linearization are mainly the following.

(1) Source degeneration resistor.
(2) Negative transconductance circuit.
(3) Using the MOS transistor working in linear region as the input stage.
(4) Floating voltage source.

Generally speaking, transconductance linearization will cause transconductance gain and equivalent transconductance decrease, which is acceptable. The following examples are given.

The gain and linearity of a simple MOS differential pair, which is not linearized, is given as shown in Fig. 4.20. And the output differential current is shown in equation (4-26).

$$i_0 = I_{d1} - I_{d2} = v_{in}\sqrt{2\beta I_s - \beta^2 v_{in}^2} \tag{4-26}$$

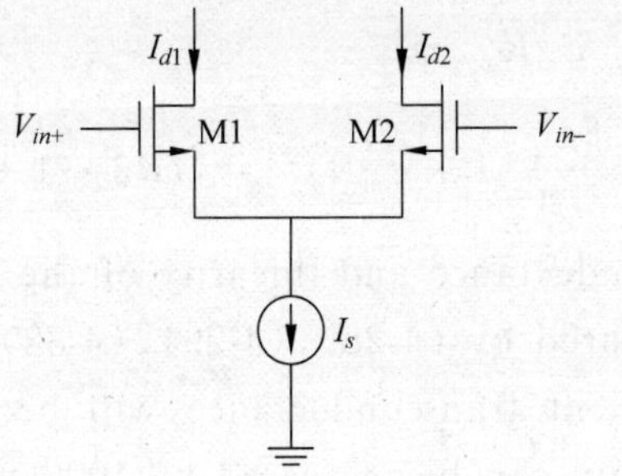

Fig. 4.20 Simple MOS differential pair

where differential input is $v_{in} = v_{in+} - v_{in-}$. Carry out Taylor expansion and take the first two items:

$$i_0 \approx \left(\frac{\partial i_0}{\partial v_{in}}\bigg|_{v_{in}=0}\right)v_{in} + \frac{1}{6}\left(\frac{\partial^3 i_0}{\partial^3 v_{in}}\bigg|_{v_{in}=0}\right)v_{in}^3 = g_m v_{in} - \frac{\beta^2}{2g_m}v_{in}^3 \tag{4-27}$$

Its equivalent transconductance g_m and linearity (input 3rd-order intercept point, IIP3) are shown as (4-28) and (4-29), respectively.

$$g_m = \sqrt{2\beta I_s} \tag{4-28}$$

$$IIP3 = \frac{8g_m^2}{3\beta^2} \tag{4-29}$$

The MOS differential pairs of resistor source degradation are shown in Fig. 4.21. and the output differential current is

$$i_0 = I_{d1} - I_{d2} = (v_{in} - i_0 R_s)\sqrt{2\beta I_s - \beta^2 (v_{in} - i_0 R_s)^2} \tag{4-30}$$

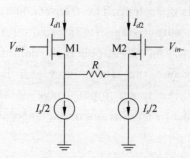

Fig. 4.21 The MOS differential pairs of resistor source degradation

Carry out Taylor expansion and take the first two items:

$$i_0 \approx \left(\frac{\partial i_0}{\partial v_{in}}\bigg|_{v_{in}=0}\right)v_{in} + \frac{1}{6}\left(\frac{\partial^3 i_0}{\partial^3 v_{in}}\bigg|_{v_{in}=0}\right)v_{in}^3 \quad (4\text{-}31)$$

$$= \frac{1}{1+g_m R_s}g_m v_{in} - \frac{1}{(1+g_m R_s)^4}\frac{\beta^2}{2g_m}v_{in}^3$$

Its equivalent transconductance G_m and $IIP3$ are shown as (4-32) and (4-33), respectively:

$$G_m = \frac{g_m}{1+g_m R_s} \quad (4\text{-}32)$$

$$IIP3_R = \frac{8g_m^2}{3\beta^2}(1+g_m R_s)^3 = IIP3 \cdot (1+g_m R_s)^3 \quad (4\text{-}33)$$

The equivalent transconductance and linearity of the differential pairs before and after linearization are compared by (4-28), (4-29), (4-32), and (4-33). Assuming that $g_m R_s = 9$, then the equivalent transconductance will become a 1/10 of that before linearization, and the linearity can be increased by 1000 times after the linearization. This price is acceptable.

4.4.4 Active resistor

Because resistor takes a lot of chip area in integrated circuit. To solve this problem, in the process, we use the diffusion method to fabricate resistor. But this resistance error is large, and it is sensitive to temperature. Because of the small area of transistors, the active devices instead of passive devices have become the first choice for designers to save the area. With the active resistor of transconductance, the high-frequency performance of circuit is improved. And the transconductance can also be adjusted to realize different resistance value. However, the nonlinearility and linear range of transconductance is small, so the linear range of resistor characteristics is limited. The active resistor realized by a transconductance is shown in Fig. 4.22.

The active resistor transfer function of the transconductance device are

$$I_{out} = -G_m V_{in} = -I_{in} \quad (4\text{-}34)$$

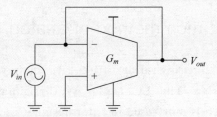

Fig. 4.22 Active resistor

$$R_{eq} = V_{in}/I_{in} = 1/G_m \tag{4-35}$$

According to (4-35), the equivalent resistance of active resistor is the reciprocal of equivalent transconductance. When the equivalent transconductance changes according to output current, the equivalent resistance value can be adjusted.

4.4.5 Active inductor

Chip area is the key problem in analog circuit, which is associated with the cost. The inductor will take up a very large area, and the inductance value can not be large, which is inconvenient for design. So the circuit designers seek other ways to replace inductors. The inductor can be designed by using the transconductance G_m and capacitor, since the transconductance G_m is adjusted continuously and easy to integrate.

The active inductor realized by G_m and capacitor is shown in Fig. 4.23, and its equivalent input impedance is expressed as (4-36).

$$Z_{in} = V_{in}/I_{in} = \frac{sC}{G_{m1}G_{m2}} \tag{4-36}$$

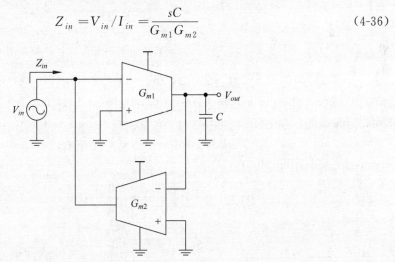

Fig. 4.23 The active inductor realized by G_m and capacitor

From (4-36), it is known that its input impedance Z_{in} presents inductor characteristics, and the equivalent inductor is:

$$L = \frac{C}{G_{m1}G_{m2}} \tag{4-37}$$

4.5 The realization method of analog filter

The methods of realizing high-order filter are mainly divided into LC ladder synthesis and cascade design. The LC ladder synthesis has the advantages of low sensitivity to element, which is more used in filter design. Cascade design method use the first order and second order filter to form poles and zeros for high-order filters. The advantages of low-order filters are the simple structure and easy to adjust. But the disadvantage is that the circuit has a high requirement for precision and stability of the elements.

4.5.1 Cascade design method

The cascade design method is based on the decomposition of the transfer function $H(s)$ of high-order filter into many low-order products. The low-order filter here can be understood as a first or second order filter. $H(s)$ can be decomposed into (4-38), as shown in Fig. 4.24.

$$H(s) = H_1(s) * H_2(s) * H_3(s) * \cdots * H_n(s) \quad (4\text{-}38)$$

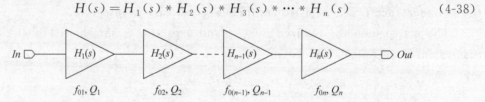

Fig. 4.24 Schematic of decomposition of high-order filters

In (4-38), $H_1(s) - H_n(s)$ are the transfer function for low-order filters. If the order L is odd, there must be a first order transfer function, while all others are second order (biquad) transfer functions. The transfer functions of cascaded filters can be expressed as (4-39) and (4-40), of which (4-39) are even-order filters, and (4-40) represent odd-order filters.

$$H_{even}(s) = \prod_i \frac{b_{2,i}s^2 + b_{1,i}s + b_{0,i}}{s^2 + \frac{\omega_{pi}}{Q_{pi}}s + \omega_{pi}^2} \quad (4\text{-}39)$$

$$H_{odd}(s) = \frac{b_1 s + b_0}{s + \omega_p} \prod_i \frac{b_{2,i}s^2 + b_{1,i}s + b_{0,i}}{s^2 + \frac{\omega_{pi}}{Q_{pi}}s + \omega_{pi}^2} \quad (4\text{-}40)$$

In implementation, if there is a first order item, then a pure RC network can be adopted, so that we only need to know its cut-off frequency. Then the second order term can get its cut-off frequency and quality factor Q through the lookup table.

Cascade design has the advantages of independent and easy adjustment. So the design is simple, and each stage can be designed and adjusted individually. Its shortcomings are also obvious, that is the independence between all stages makes the overall filter sensitive to the components parameter.

Mathematically speaking, the order of each part is independent. But in practical applications, the low-order filter must be connected in a certain order to maximize the performance. Due to the presence of signal clamping in second order with high quality factor, in order to avoid loss of dynamic range and precision reduction, the biquads are connected in series in order of Q values from low to high. This order can also minimize the noise of filter. Table 4-1 is the normalized coefficient of different order Butterworth filters with cascade design method.

Table4-1　the normalized coefficient of different order Butterworth filters with cascade design method

	Butterworth Low-Pass Filter									
N	f_{01}	Q_1	f_{02}	Q_2	f_{03}	Q_3	f_{04}	Q_4	f_{05}	Q_5	$2f_c$ attenuation
2	1	0.707	1								12.30
3	1	1.000	1								18.13
4	1	0.541	1	1.306							24.10
5	1	0.618	1	1.620	1						30.11
6	1	0.518	1	0.707	1	1.932					36.12
7	1	0.555	1	0.802	1	2.247	1				42.14
8	1	0.510	1	0.601	1	0.900	1	2.563			48.16
9	1	0.532	1	0.653	1	1.000	1	2.879	1		54.19
10	1	0.506	1	0.561	1	0.707	1	1.101	1	3.196	60.21

4.5.2　LC ladder synthesis

LC ladder synthesis is to use the whole method to design filter. The difference between this method and the cascade design method is that the filter is a tight coupling system which is considered to be the lowest sensitivity to the components. Because this sensitivity is distributed to all of its components, rather than on any independent device.

Fig. 4.25 gives an implementation of LC ladder synthesis circuit. Vs is an AC voltage source. Rs is the resistor of AC voltage source. R_{out} is output resistor, and V_{out} is the output voltage. $L_1 - L_n$ are inductors, and $C_1 - C_{n-1}$ are capacitors.

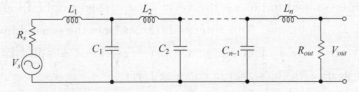

Fig. 4.25　Realization of LC ladder synthesis

In low frequency, the inductor is approximately short, and the capacitors is approximately open. A signal path is formed from the input to output. The voltage gain is $R_{out}/(R_{out}+R_s)$. In the medium frequency band, due to the existence of a series resonant circuit, the response presents a series of notch, and the response order n is the sum of inductors and capacitors number. Fig. 4.25 shows a full-pole ladder network circuit. The transfer function of this circuit has no zeroes, and can be used to compose Butterworth, Chebyshev and Bessel response. Table 4-2 is the Butterworth polynomial approximation of a LC ladder network.

Table4-2 Butterworth polynomial approximation of a LC ladder network

n	Butterworth polynomial approximation
1	$s+1$
2	$s^2+1.4142s+1$
3	$(s+1)(s^2+s+1)$
4	$(s^2+0.765s+1)(s^2+1.848s+1)$
5	$(s+1)(s^2+0.618s+1)(s^2+1.618s+1)$
6	$(s^2+0.5176s+1)(s^2+1.4142s+1)(s^2+1.9318s+1)$
7	$(s+1)(s^2+0.445s+1)(s^2+1.247s+1)(s^2+1.802s+1)$
8	$(s^2+0.3902s+1)(s^2+1.1112s+1)(s^2+1.663s+1)(s^2+1.9616s+1)$
9	$(s+1)(s^2+0.3474s+1)(s^2+s+1)(s^2+1.532s+1)(s^2+1.8794s+1)$
10	$(s^2+0.3128s+1)(s^2+0.908s+1)(s^2+1.4142s+1)(s^2+1.782s+1)(s^2+1.9754s+1)$

4.6 Complex filter

In a RF receiver with low frequency (or Intermediate Frequency, IF) structure, a designer often uses a complex filter to eliminate the effect of mirror signal, as shown in Fig. 4.26. After the down conversion of RF signal(A) through the orthogonal local oscillator signal(B), the desired signal and the mirror signal will split up in frequency domain, respectively, on both sides of the imaginary axis (C). If a non-imaginary axisymmetric filter is used to process them, the mirror signal can be filtered out and the required signal(D) is obtained. This filter is different from the general low-pass or high-pass filter (they are all imaginary axisymmetric). It is called complex filter, because in time domain it deals with the signal in complex form.

The complex filter is also divided into two types: passive and active. The passive complex filter is implemented by RC-CR phase-shifting network. Its advantages are

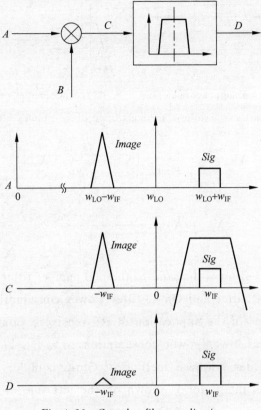

Fig. 4.26 Complex filter application

simple structure and no power consumption, but the drawback is that it will increase signal loss and do not have channel selection function. While the active complex filter can also play the role of channel selection while eliminating the mirror signal. As shown in Fig. 4.27(a), it is a common method to use the low-pass filter prototype to get the active complex filter through frequency shift. Taking the first order low-pass filter as an example, its transfer function is

$$H(s) = \frac{\omega_0}{S + \omega_{LP}} \tag{4-41}$$

It is about the imaginary axisymmetric, and the transfer function of the first order complex filter is obtained after frequency shift of ω_{IF}.

$$H(s)_{CF} = \frac{\omega_0}{(S - j\omega_{IF}) + \omega_{LP}} \tag{4-42}$$

After the frequency shift, the transfer function is $j\omega_{IF}$ symmetric.

The transformation principle from low-pass prototype to complex filter is shown in Fig. 4.27, which outputs X_{out} multiplied by the coefficient $j\omega_{IF}/\omega_0$ and returns it to the input

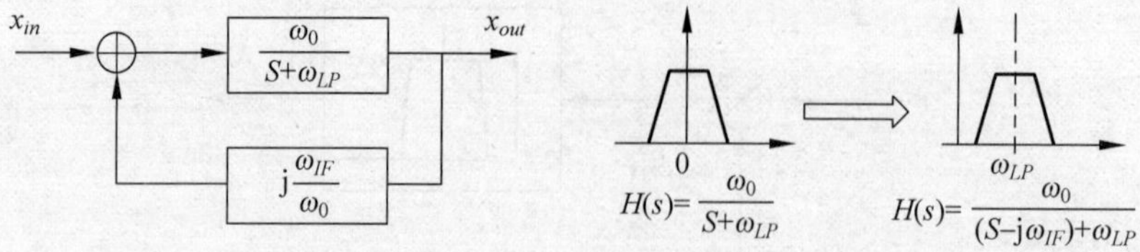

(a) The transfer function of a complex filter (b) The basic principle of complex filter

Fig. 4.27 Synthesis method of first order complex filter

$$X_{out} = \frac{\omega_0}{S+\omega_{LP}} \cdot \left(X_{in} + j\frac{\omega_{IF}}{j\omega_{IF}} \cdot X_{out}\right) \quad (4\text{-}43)$$

then:

$$X_{out} = X_{in} \cdot \frac{\omega_0}{\left(S - j\frac{\omega_{IF}}{\omega_0}\right) + \omega_{LP}} \quad (4\text{-}44)$$

The active complex filter is the same as the real number filter, which can be realized in two ways: active RC filter and Gm-C filter. Power consumption is one of the most important considerations in the application of RF receivers. Compared with the active RC, the Gm-C filter has lower power consumption, so it has been used more widely. Fig. 4.28 is the first order synthesis method of Gm-C complex filter. The first order single-end output Gm-C integrator is shown on the left side of Figure 4.28(a). The transconductance value is $\omega_0 C$, and the relationship between input and output is as follows:

$$X_{oI} = X_{iI} \cdot \frac{\omega_0 C}{sC} = X_{iI} \cdot \frac{\omega_0}{s} \quad (4\text{-}45)$$

$$X_{oQ} = X_{iQ} \cdot \frac{\omega_0 C}{sC} = X_{iQ} \cdot \frac{\omega_0}{s} \quad (4\text{-}46)$$

In order to achieve frequency shift, transconductance unit $\omega_{IF}C$ and $-\omega_{IF}C$ are connected to each output of the two I/Q outputs, and two outputs are:

$$X_{oI} = (X_{iI} \cdot \omega_0 C - X_{oQ} \cdot \omega_{IF} C) \cdot \frac{1}{sC} \quad (4\text{-}47)$$

$$X_{oQ} = (X_{iQ} \cdot \omega_0 C + X_{oI} \cdot \omega_{IF} C) \cdot \frac{1}{sC} \quad (4\text{-}48)$$

With the addition of two outputs, the final single-end output can be obtained as follows:

$$X_o = X_{oI} + jX_{oQ} = \frac{\omega_0}{s - j \cdot \omega_{IF}} \cdot X_i \quad (4\text{-}49)$$

As a differential mode, the circuit structure is shown in Fig. 4.28(b).

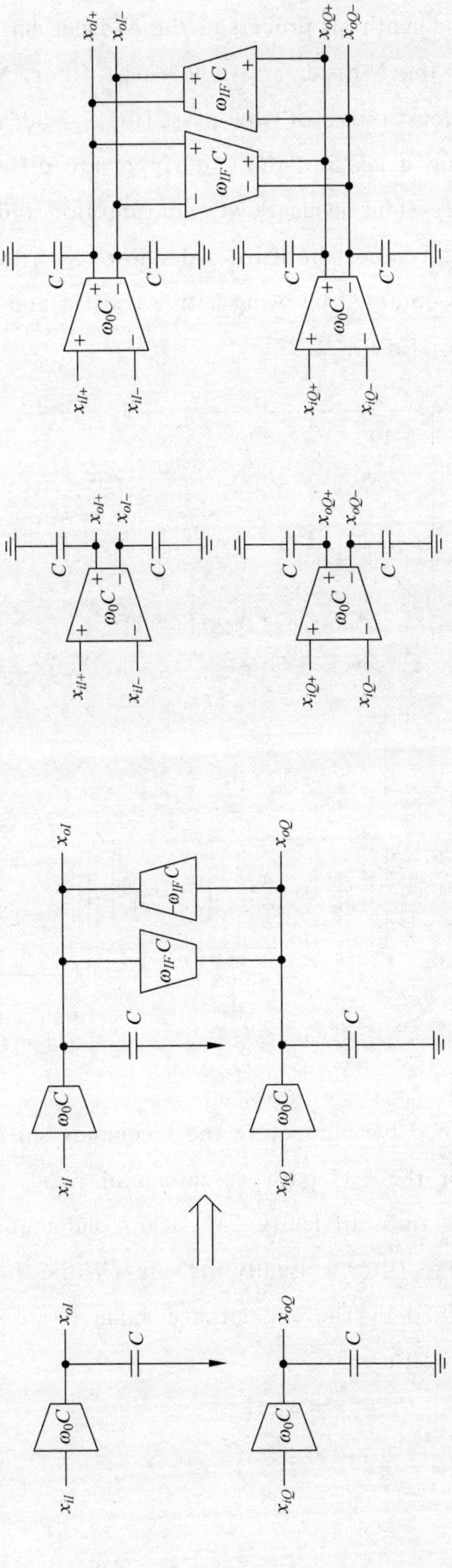

Fig. 4.28 Gm-C synthesis method of first order complex filter

(a) single-end implementation

(b) differential implementation

In design, a high-order complex filter is usually needed to meet the Image Rejection Ratio(IMRR) requirements. The synthesis process of the N-order active complex filter is obtained by frequency shift of the N-order active low-pass filter. As described in the above sections, there are various types of low-pass filters, and their out of band attenuation characteristics, group delay and in-band ripple are different. They can be selected according to the actual system needs. Power consumption and difficulty are also considered. Fig. 4.29 is the prototype of the third-order low-pass filter. The Gm-C low-pass filter shown in Fig. 4.30 is obtained by using active resistor and active inductors to replace the resistors and inductors in Fig. 4.29.

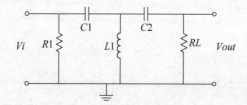

Fig. 4.29 prototype of the third-order low-pass filter

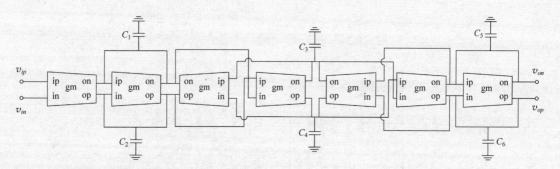

Fig. 4.30 Gm-C low-pass filter

The complex filter is obtained by connecting the frequency shift transconductance between each capacitor node of the I/Q path, as shown in Figure 4.31. In practical design, in order to reduce the difficulty of transconductance design, every transconductance value in low-pass filter is usually the same. While the transconductance with frequency shift is determined by the capacitance value C_i of the node and the center frequency ω_{IF} of complex filter.

$$g_{mi} = \omega_{IF} C_i \tag{4-50}$$

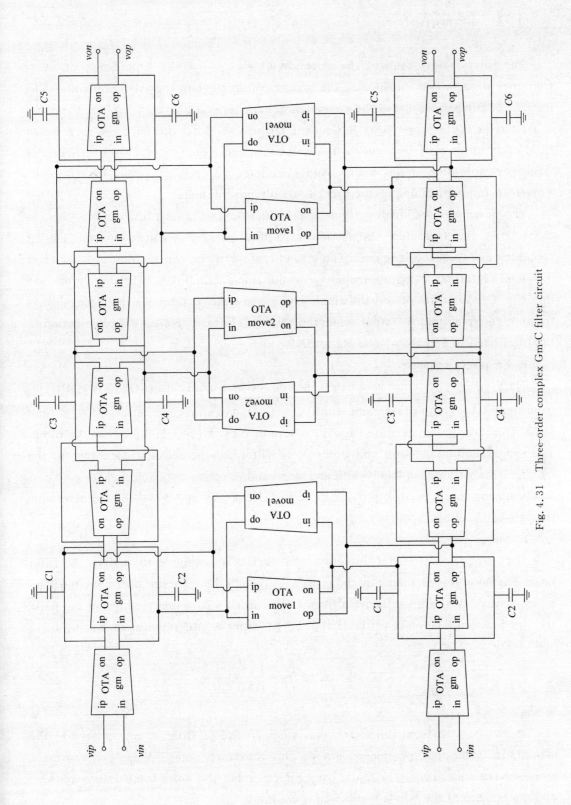

Fig. 4.31 Three-order complex Gm–C filter circuit

4.7 Parameter

The filter mainly realizes the selection of signal and the suppression of other frequency signal, which can reduce the system requirement to the ADC resolution. The increase of filter order makes the group delay characteristic become worse and the phase characteristic is the worst near the cut-off frequency. Moreover, with the increase of filter order, the cut-off frequency is not well controlled, which increases the difficulty of frequency calibration circuit design. And sometimes we need to correct the Q value. Therefore, in practical design, the order is usually not too high.

The parameters of analog filter mainly include two kinds: the first is used to characterize the frequency suppression capability parameters, including the cut-off frequency and frequency response. The second represents the contribution of filter error to system performance, mainly including inband ripple, group delay variation, noise and linearity. The variation of in-band ripple and group delay determines the signal quality transmitted from input to output, while the noise and linearity reflect the contribution of the filter itself to Signal-to-Noise Ratio(SNR) loss.

1. Frequency response

The frequency response parameters include the pass-band cut-off frequency(ω_p), in-band ripple(A_{PB}), transition band width($\omega_s - \omega_p$), and stop-band rejection($A_{\min}$) and so on. The three parameters are basically determined by the system directly to the filter. The approximation function and order L of filter can be selected according to the requirement of transition band width and stop-band rejection parameters. And band-pass and band-stop filters need to provide the center frequency and bandwidth to determine the pass-band cut-off frequency.

2. Noise

The first order RC filter(integrator) is used as an example to analyze the filter noise. The noise is contributed by the resistor. Although the capacitor does not contribute to noise, it connects with resistor to form a first order zero, which will affect the noise transfer function. The noise of the first order RC filter is integrated, as shown in (4-51) and (4-52).

$$\overline{V_n^2} = \int_0^\infty 4KTR \left| \frac{1}{1+j\omega RC} \right|^2 df \qquad (4\text{-}51)$$

$$\overline{V_n^2} = KT/C \qquad (4\text{-}52)$$

As can be seen from the above, the integral noise of filter is only related to the capacitance value, but is independent of the resistance value. When the resistance increases, the noise density in signal band increases, but the noise bandwidth is reduced and the integral of the whole bandwidth is constant.

3. Dynamic range

The Dynamic Range(DR) of the filter is defined as the ratio of the maximum signal to the minimum signal that the filter can handle. The ability to deal with the maximum signal is determined by the nonlinear and output swing of circuit. While the ability to process the minimum signal is usually determined by noise. Because the signal is too weak to be "drowned" in the noise. Assuming that the maximum voltage swing can reach the power supply voltage V_{dd}, then for a sinusoidal signal, the maximum signal amplitude is (4-53), and the minimum signal amplitude is (4-54).

$$V_{\max}(\text{rms}) = \frac{V_{dd}}{2\sqrt{2}} \tag{4-53}$$

$$V_{\min}(\text{rms}) = \beta\sqrt{KT/C} \tag{4-54}$$

where β represents high-order filter noise coefficient. From (4-53) and (4-54), DR can be obtained as

$$DR(\text{dB}) = 20\log_{10}\frac{V_{\max}(\text{rms})}{V_{\min}(\text{rms})} = 20\log_{10}\frac{V_{dd}}{2\beta\sqrt{2KT/C}} \tag{4-55}$$

4.8 Technical words and phrases

4.8.1 Terminology

passive filter	无源滤波器
active filter	有源滤波器
low-pass filter	低通滤波器
high-pass filter	高通滤波器
band-pass filter	带通滤波器
band-stop filter	带阻滤波器
all-pass filter	全通滤波器
Butterworth filter	巴特沃兹滤波器
Chebyshev filter	切比雪夫滤波器
Bessel filter	贝塞尔滤波器
elliptic filter	椭圆滤波器
ripple	纹波
group delay	群时延
amplitude-frequency characteristic	幅频特性
transfer function	传递函数
biquad	二阶节
active RC integrator	有源 RC 滤波器
source degeneration resistor	源退化电阻
LC ladder synthesis	LC 梯形综合设计法
complex filter	复数滤波器
Image Rejection Ratio(IMRR)	镜像抑制比
dynamic range(DR)	动态范围

4.8.2　Note to the text

(1) In the light of approximation methods, it consists of Butterworth, Chebyshev, Bessel and elliptic filter.
根据不同的滤波器逼近方式,可分为巴特沃兹、切比雪夫、贝塞尔和椭圆滤波器等。

(2) Active filter has the advantages of low noise and small area.
有源滤波器具有噪声系数低、占用芯片面积小等优点。

(3) The integrator, realized by resistor, capacitor and OTA, is called an active *RC* integrator.
采用电阻、电容和运算放大器实现的积分器称为有源 *RC* 积分器。

(4) Because the MOS transistor operating in linear region can be used as a resistor, and the resistance can be continuously adjustable according to gate voltage. So the application of this adjustable resistor in active *RC* integrator can improve its frequency accuracy.
由于工作在线性区的 MOS 晶体管可以作为电阻,并且阻值可以根据晶体管的栅源电压连续可调,因此将这种可调电阻应用在有源 *RC* 积分器中可以提高其频率精度。

(5) When an OTA is used as an active device, its limited gain and bandwidth seriously restrict the operating frequency of filter. Therefore, when frequency is high, the transconductance device is used as an active device. The integrator, consisting of a transconductance and capacitor, is called a *Gm-C* integrator.
当采用运算放大器作为有源器件时,其有限增益和带宽严重限制了滤波器的工作频率。所以当处理的信号频率较高时,采用跨导器作为有源器件。由跨导器和电容构成的积分器称为 *Gm-C* 积分器。

(6) Because resistor takes a lot of chip area in integrated circuit. To solve this problem, in the process, we use the diffusion method to fabricate resistor. But this resistance error is large, and it is sensitive to temperature.
由于电阻在集成电路中需要占用大量的芯片面积,为解决这个问题,在工艺上有利用扩散法制造电阻,只是这种电阻误差大,对温度比较敏感。

(7) Cascade design method use the first order and second order filter to form poles and zeros for high-order filters.
级联设计法是采用一阶和二阶滤波器级联,形成高阶滤波器所需要的极点和零点。

Chapter 5 Comparator

CHAPTER 5

With the rapid development of integrated circuit, the mixed-signal integrated system, especially the VLSI chip represented by System-On-Chip (SoC), has a large number of analog-to-digital converters, automatic gain control loops (AGC), peak detectors and other circuits. As a key module in these circuits, comparators have been an important circuit module in academia and industry. Their speed, power consumption, noise, offset voltage and other performances play a crucial role in the speed, accuracy and power consumption of the whole system.

5.1 Basis of comparator

The main function of comparator circuit is to compare an analog signal with another or reference signal, and output it to get the high and low voltage as binary signal through comparison processing. In the ideal case, when the difference between positive and negative inputs of comparator is positive, the output is high (V_{OH}). And when the input difference is negative, the comparator output is low (V_{OL}). The ideal transfer curve of comparator is shown in Fig. 5.1, where V_p is the inverse input, V_n is the inverse input. And the maximum and minimum value of comparator output is defined as V_{OH} and V_{OL} respectively. In practical circuit, V_{OH} and V_{OL} usually correspond to the power supply voltage and the ground voltage respectively.

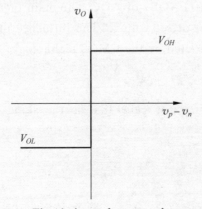

Fig. 5.1 The ideal transfer curve of comparator

When the voltage difference between two input is zero, the comparator output will not change. Actually comparators can not identify the tiny voltage difference without limitation. Due to the limitation of finite gain, there is usually a minimum resolvable voltage difference, which is called the accuracy(or resolution) of comparators. Figure 5.2 shows the transfer curve of a finite-gain comparator.

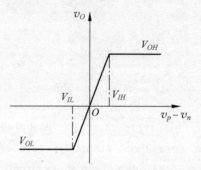

Fig. 5.2 The transfer curve of a finite-gain comparator

The V_{IH} and V_{IL} are the input voltage difference $V_p - V_n$ required by the output to reach the upper limit and the lower limit, which is the accuracy (resolution) of the comparator.

5.2 Parameter

The comparator parameters include two aspects of static and dynamic characteristics. The static characteristics include gain, resolution, offset voltage, etc. The dynamic characteristics mainly include the operating characteristics of small and large signals. The specific definitions of the various parameters are given below.

1. Resolution

Resolution is the minimum input voltage difference that can produce correct digital output. In some analog-to-digital converters, such as Flash ADC and successive approximation analog-to-digital converter, the comparator resolution directly determines the Effective Number Of Bit(ENOB) of ADC. The main factors affecting the resolution are noise, gain and the input offset voltage. The influence of offset voltage is the most serious, and it is mainly restricted by the CMOS technology. The definition is expressed:

$$\Delta V = \frac{V_{OH} - V_{OL}}{A_v} \tag{5-1}$$

where A_v is the comparator gain, that is the slope of transition curve, whose expression is

$$A_v = \frac{V_{OH} - V_{OL}}{V_{IH} - V_{IL}} \tag{5-2}$$

2. Delay

The delay is generally defined as the time difference between the input analog signal and the output digital signal. This parameter determines the maximum operating frequency of comparator.

3. Slew rate

The delay of comparator varies with the change of input amplitude, and the larger input will make the delay shorter. When the input is increased to an upper limit that will not affect the delay, the voltage change rate is called slew rate.

4. Kick-back noise

Kick-back noise refers to the noise generated by digital output to the input analog signal. The noise is usually caused by charge feedback.

5. V_{offset} (V_{OS})

The input offset voltage is produced by the mismatch of input differential MOS transistors or process. The MOS device shows a more serious input offset voltage than the BJT transistor. And the input offset voltage is also an important factor affecting the resolution of comparator. It is defined as: if the two input of comparator are connected to the same voltage value, the output is the offset voltage. The transfer curve of comparator introduced into the input offset voltage is shown in Fig. 5.3.

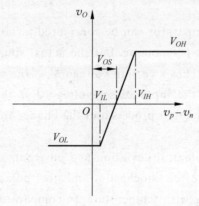

Fig. 5.3 The transfer curve of comparator introduced into the input offset voltage

6. Input common mode range

The input common mode range is the difference range of input voltage that the comparator can continuously distinguish. This characteristic is also one of the important characteristics of comparator.

7. Differential input voltage range

The differential input voltage range is defined as the maximum voltage allowed by the two input of comparator.

8. Output swing

The comparator outputs positive voltage when the inphase input voltage is greater than the negative input. The output swing is determined by the differential amplifier and the bias network within comparator, and it is also affected by the power supply voltage.

5.3　Characteristic analysis

The analysis of the comparator mainly divided into static and dynamic characteristic analysis. The following is the specific analysis.

1. Static characteristics analysis

The gain of the actual comparator is defined as $A_v = (V_{OH} - V_{OL})/V_{IH} - V_{IL}$, which is a finite value. V_{IH} and V_{IL} are the inphase and inverse input voltages required for the output to reach the upper and the lower limit. V_{OH} and V_{OL} are the output high and low voltage respectively. The gain is usually considered a function of its input signal. In structure, there are a variety of ways to improve its gain value. It is commonly used to add a or multi-stage pre-amplifier before the comparator; and add a inverter after the comparator to pull the output to the supply voltage or ground.

The resolution is the difference between the smallest input signal that the comparator can distinguish. It can be seen that the relationship between the resolution and the gain is very close, and the high resolution comparator circuit also means that its gain is higher.

The gain of an ideal comparator can be considered infinitely large, that is, when the input crosses zero, the output is changing. But the actual situation is that only when the input differential voltage reaches a certain voltage V_{OS}, the output starts to change. The voltage V_{OS} at this time is the input offset voltage. For the input offset voltage, the mismatch introduced in production process and the change of environment are the main reasons.

For the offset of technological deviation and environmental change, the magnitude of input offset voltage is often stochastic, and the polarity of its voltage is also unpredictable and drift along with temperature. In comparator circuit design, the impact of the offset voltage can be reduced by introducing an input or output offset storage technology.

In the common mode input range, the comparator input can be processed to output the right digital code, that is, the input transistors are in the normal operating state. At this point, the resolution and the input offset voltage can be considered as a function of the input common mode range.

2. Dynamic analysis

To analyze the dynamic characteristics, the small-signal and the large-signal characteristics of comparator will be discussed. First, when the input signal is small, the analysis is done by the small-signal analysis method. While the input signal increases, the delay decreases. Until the input amplitude increases to a certain extent, even if the input signal continues to increase, the delay will no longer change. The voltage change rate at this time is called Slew Rate (SR). As the input signal continues to increase, the comparator eventually enters the large-signal mode. In these two operating mode, the determinants of comparator's delay are different. In general analysis, in small-signal mode, the larger input amplitude and higher gain will shorten the delay time.

For small-signal behavior, the delay is mainly caused by the nonlinear characteristics of circuit. For large-signal behavior, SR is mainly limited by the output driving capacity, which is shown as the charge and discharge speed of load capacitor. In comparator design, if the delay of jitter is required smaller, we should make SR a major determinant, and avoid the influence of zero/pole in signal frequency range. Then the delay can be expressed as

$$\tau_p = \frac{V_{OH} - V_{OL}}{2 \cdot SR} \tag{5-3}$$

Therefore, in order to reduce the delay, it is necessary to increase the current capacity and SR of comparator. It also means that there is a certain tradeoff between the power and speed.

5.4 Comparator structure

From the principle of operation, all comparators can be regarded as the different applications of amplifier. So it can be divided into two basic structures: open-loop and closed-loop. A high-gain OPA in open-loop state is a high-resolution comparator, and the hysteresis comparator and latch circuit is closed-loop amplifier with positive feedback.

From the aspect of power consumption, comparator includes static and dynamic comparator. The main difference between the two is that the static comparator consumes a certain static power. While the static power consumption of dynamic comparator is zero, and it only consumes the dynamic power dissipation.

According to the principle of operation, the comparator circuit can also be divided into the open-loop comparator and the regeneration comparator. Based on circuit structure, it consists of single-end output and differential-output structure. In design, it is more reasonable to choose the corresponding comparator circuit structure according to the application scene. The following is a brief introduction to open-loop comparator and dynamic comparator.

1. Open-loop comparator

The open-loop comparator is realized by open-loop amplifier. Such comparators do not need frequency compensation, so that the maximum bandwidth can be obtained. Meanwhile in theory, a relatively fast response time can be acquired. This comparator can be classified into a single-stage high-gain comparator and a low-gain multi-stage cascade comparator according to the amplifier structure.

The comparator formed by open-loop single-stage amplifier mainly depends on the amplifier's high-gain to enlarge the input differential signal to supply voltage and ground, so as to output digital code "1" and "0". This comparator does not have feedback and circuit structure is simple. However, it can not be used in high-resolution systems because of its poor performance of offset voltage, setup time, and slew rate. And because the DC gain is high, and the bandwidth is small, the setup time is relatively long, which is generally suitable for single-pole system and small-signal input application.

Considering the setup time, if we want to increase the comparison speed, we need to increase the dominant pole frequency of amplifier and ensure its original unit gain bandwidth unchanged. This method usually sacrifices a certain DC gain. In order to compensate for the decrease of DC gain, multi low-gain amplifiers can be cascaded to form a comparator circuit.

2. Dynamic comparator

The dynamic comparator is mainly divided into the resistor-divider comparator, the differential-pairs comparator and the charge-distribution comparator. Other kinds of comparators are usually improved on the basis of these comparators.

The structure of resistor-divider comparator is shown in Fig.5.4. The transistors M1~M4 operates in the linear region, which is equivalent to the voltage-control resistor. It can adjust the threshold voltage of comparator by changing its resistance value, and the M5~M12 forms the latch. Assuming the length of M1~M4 are the same, and $W_A = W_2 = W_4$, $W_B = W_1 = W_3$, the threshold of comparator is

$$V_{in}^+ - V_{in}^- = \frac{W_B}{W_A}(V_{ref}^+ - V_{ref}^-) \tag{5-4}$$

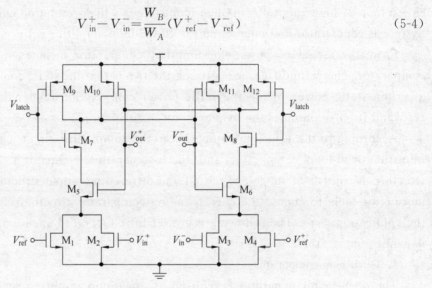

Fig. 5.4 The resistor-divider comparator

In this structure, the offset is mainly affected by M1~M4, and the effect of M5 and M6 on offset is relatively small. Because of the smaller size of transistors and the input transistors which have a large impact on offset operating in linear region, the structure has a larger offset. At the same time, since the large output changes has less effect on M1~M4 drain, this circuit is with low kick-back noise.

The differential-pairs comparator structure, as shown in Fig.5.5, is composed of two cross-coupled differential pairs with a switch-controlled current source and a latch. The threshold can be set by introducing the imbalance of coupled pairs. Assuming that the length of M1 ~ M4 are equal and $W_1 = W_2$, $W_3 = W_4$, then the current of coupled pairs is expressed as

$$I_{D1} - I_{D2} = \beta_1 V_{in} \sqrt{\frac{2I_{D5}}{\beta_1} - V_{in}^2} \tag{5-5}$$

$$I_{D4} - I_{D3} = \beta_3 V_{in} \sqrt{\frac{2I_{D6}}{\beta_3} - V_{ref}^2} \tag{5-6}$$

where $\beta_i = (1/2)\mu C_{ox}(W_i/L)$, $V_{in} = V_{in}^+ - V_{in}^-$, $V_{ref} = V_{ref}^+ - V_{ref}^-$. When $I_1 = I_{D1} + I_{D3}$ equals $I_2 = I_{D2} + I_{D4}$, The state of comparator changes.

The offset voltage is mainly influenced by M1~M4. But its offset is smaller than the resistor-divider comparator. Unlike resistor-divider comparator, the large output changes has more effect on M1~M4 drain, so it generates high kick-back noise.

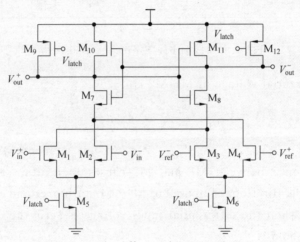

Fig. 5.5 The differential-pairs comparator

The charge-distribution comparator is demonstrated in Fig.5.6. Its operating principle is as follows: when $\overline{latch}$ is high, the gate of M1 and M2 is connected to ground. C_{in} and C_{ref} are charged by V_{in}^- and V_{ref}^- respectively. At this time, the charge in C_{in} is $Q_{in} = V_{in} \cdot C_{in}$, and C_{ref} is $Q_{ref} = V_{ref} \cdot C_{ref}$. Since latch is low, M3 is cut-off, and M6、M9 turns on, which set V_{out}^- and V_{out}^+ as VDD and the latch circuit keeps the value of last judge. Meanwhile, M4 and M5 turns on that connects the drain of M1、M2 to VDD. After that latch changes to high, the bottom plate of C_{in} and C_{ref} connects to ground. According to the principle of conservation of charge:

$$V_{in}^- \cdot C_{in} + V_{ref}^- \cdot C_{ref} = V_- \cdot (C_{in} + C_{ref}) \tag{5-7}$$

$$V^- = V_{in}^- \frac{C_{in}}{C_{in} + C_{ref}} + V_{ref}^- \frac{C_{ref}}{C_{in} + C_{ref}} \tag{5-8}$$

When latch is high, the gate voltage of M1 is:

$$V^+ = V_{in}^+ \frac{C_{in}}{C_{in} + C_{ref}} + V_{ref}^+ \frac{C_{ref}}{C_{in} + C_{ref}} \tag{5-9}$$

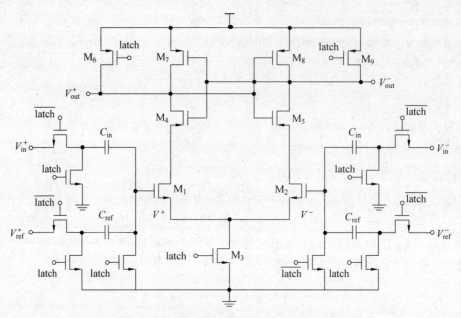

Fig. 5.6 The charge-distribution comparator

the input differential voltage of M1, M2 is:

$$V = V_{in} \frac{C_{in}}{C_{in} + C_{ref}} - V_{ref} \frac{C_{ref}}{C_{in} + C_{ref}} \tag{5-10}$$

where $V_{in} = V_{in}^+ - V_{in}^-$, $V_{ref} = V_{ref}^+ - V_{ref}^-$. At this moment, M3 turns on, and the gate voltage difference between M1 and M2 causes their drain to generate voltage difference, so that the flip flops composed of M4~M7 turn over and output signals. From the above analysis, when the differential input voltage is zero, the comparator start to work, thus the threshold is

$$V_{TH} = -V_{ref} \frac{C_{ref}}{C_{in}} \tag{5-11}$$

By setting the ratio of C_{in} and C_{ref}, the threshold of comparator can be adjusted. The offset of the structure is mainly determined by the mismatch of input differential pairs and the deviation of capacitance ratio. Besides kick-back noise, the error in the input switches is caused by charge injection.

The performance of these three kinds of dynamic comparator circuits is compared in Table 5.1.

Table5.1 Performance comparison

Category	Offset	Kick-back noise	Speed	Area	Power
the resistor-divider	large	small	slow	small	small
The differential-pairs	small	large	fast	medium	medium
The charge-distribution	small	medium	large	large	large

5.5　Basis of schmitt trigger

Schmitt trigger is actually a special comparator circuit that contains positive feedback. For standard Schmitt trigger, when the input voltage is higher than the forward threshold, the output is high; when the input voltage is below the negative threshold, the output is low; while the input is between positive and negative threshold, the output will not change. That is, when the output turns from high to low, or from low to high, the threshold is different. The output will change only when the input voltage varies enough, so this circuit is named as a trigger. This double-threshold action is called hysteresis, indicating the memory of Schmidt trigger. In essence, Schmitt trigger can be considered a bistable multivibrator.

Schmitt trigger can be used as a waveform shaping circuit to shape the analog signal waveform into square wave that the digital circuit can handle. Moreover, because Schmitt trigger has hysteretic characteristics, its applications include anti-disturbance in open-loop configuration, and realization of multivibrator in closed-loop positive feedback configuration. Schmidt trigger is divided into two kinds: non-reverse Schmidt trigger and reverse Schmidt trigger. The symbol and voltage transmission characteristics are shown in Fig.5.7(a) and (b).

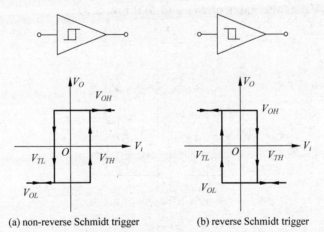

(a) non-reverse Schmidt trigger　　(b) reverse Schmidt trigger

Fig. 5.7　The symbol and voltage transmission characteristics of two kinds of Schmidt trigger

In the discrete device system, Schmidt trigger circuit is usually built by an operational amplifier. In CMOS integrated circuits, Schmidt trigger are made up of only a few NMOS and PMOS transistors, a typical Schmidt trigger is shown in Fig.5.8. The Schmidt trigger contains two similar sub-circuit structures (M1, M2, M3 and M4, M5, M6). Each sub-circuit can be regarded as a nonlinear load of another sub-circuit. But when it is in the working state, that is, at the transition point, each of the sub-circuits can be regarded as the linear resistor load of the other sub-circuit.

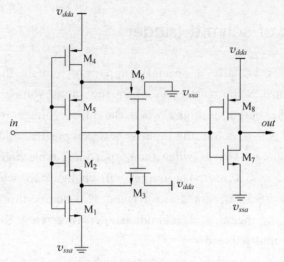

Fig. 5.8 Schmidt trigger

1. The current-voltage characteristics

In the circuit of Fig. 5.8, the bottom circuit M1, M2, M3 (which is called here the N-subcircuit), is loaded by the top circuit, M4, M5, M6 (P-subcircuit). To obtain the voltage-current characteristics of these nonlinear loads, one can take, for example, the N-subcircuit, apply a voltage source V_0, and calculate the source current I_0, assuming a constant voltage V_G at the gates of M1 and M2 (Fig. 5.9)

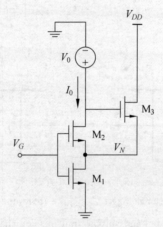

Fig. 5.9 Half equivalent circuit

When the voltage V_0 is very small, transistor M3 will be off, and M1 and M2 are in the triode mode of operation. The current I_0 is equal to

$$I_0 = 2k_1(V_G - V_{TN})V_N \tag{5-12}$$

where V_{TN} is the threshold of NMOS, $V_G - V_{TN}$ is the overdrive voltage. if one considers transistor M1, or:

$$I_0 = 2k_2(V_G - V_N - V_{TN})(V_0 - V_N) \tag{5-13}$$

In (5-12) and (5-13), it is assumed that $V_G > V_{TN}$, For the triode mode of

operation, $V_N \ll V_{TN}$, and (5-13) can be simplified to:
$$I_0 = 2k_2(V_G - V_{TN})(V_0 - V_{TN}) \tag{5-14}$$
so from (5-12) and (5-14), we can get:
$$V_N = V_0 \cdot \frac{k_2}{k_1 + k_2} \tag{5-15}$$
$$I_0 = \frac{2k_1 k_2 (V_G - V_{TN})}{k_1 + k_2} \cdot V_0 \tag{5-16}$$
from (5-16) the equivalent resistance is:
$$R_{LN} = \left[\frac{\partial I_0}{\partial V_0}\right]^{-1} = \frac{k_1^{-1} + k_2^{-1}}{2(V_G - V_{TN})} \tag{5-17}$$

It is seen from (5-15) and (5-17) that, in this part of the subcircuit operation, transistors M1 and M2 may be considered as a series connection of two resistors.

When V_0 increases, M2 enters into saturation. Then I_0 is determined, depending on the considered transistor, or by
$$I_0 = 2k_1[V_G - V_{TN} - (V_N/2)]V_N \tag{5-18}$$
or:
$$I_0 = k_2(V_G - V_N - V_{TN})^2 \tag{5-19}$$
from (5-18) and (5-19) one can find that
$$V_N = (V_G - V_{TN})\left(1 - \sqrt{\frac{k_1}{k_1 + k_2}}\right) \tag{5-20}$$
from (5-20) when the voltage V_0 achieves the value of $V_{0s} = V_G - V_{TN}$, the current I_0 becomes constant and equal to
$$I_{0N} = \frac{k_1 k_2}{k_1 + k_2}(V_G - V_{TN})^2 \tag{5-21}$$
Yet, an additional increase of V_0 will gradually introduce some changes. When V_0 achieves the value of
$$V_{0T} = V_G - (V_G - V_{TN})\sqrt{\frac{k_1}{k_1 + k_2}} \tag{5-22}$$
Then transistor M3 will be turned on, V_N starts to increase again, and the current I_0 is diminishing. When V_0 becomes equal to
$$V_{0c} = V_G + (V_G - V_{TN})\sqrt{k_1/k_3} \tag{5-23}$$
transistor M2 will be completely turned off and I_0 becomes equal to zero. At this instant, voltage V_N will be equal to $V_N = V_G - V_{TN}$, and M1 is entering into saturation. Transistor M1 carries the current
$$I_N = k_1(V_G - V_{TN})^2 \tag{5-24}$$
which is completely intercepted by M3. Additional increase of V_0 up to V_{DD} does not bring any changes and completes the current-voltage characteristic of the N-subcircuit. The analysis of current-voltage characteristics of the N-subcircuit is shown in Fig. 5.10.

Now the design problem can be formulated graphically [Fig. 5.11]. Assuming that the trigger transition from one stable state to another takes place when the gate voltage has a required threshold value V_H, and allowing a current ΔI to flow at this instant in the transistors M1, M2, M3, and M4, one has to find and superimpose the current-voltage characteristics of the two subcircuits so that only one unstable intersection point exists. A similar condition is then applied for another transition point, characterized by another required threshold voltage V_L. The characteristics shown in Fig. 5.10 help to analyze the trigger behavior near the transition point and to apply it to the circuit of Fig. 5.8.

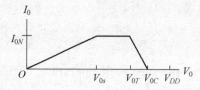

Fig. 5.10 current-voltage characteristics of the N-subcircuit

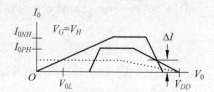

Fig. 5.11 The design problem

2. Threshold design

First it is assuming that the input $V_G = 0$ in Fig. 5.9. Then transistors M1 and M2 are off. Transistors M4 and Ms are in the linear mode of operation, but the voltage drop at each is zero because the current in M4 and Ms is equal to the current in M1 and M2. The output voltage V_0 is equal to V_{dd}. Transistor M3 is on (its drain and gate have the same voltage of V_{dd}) but it also does not carry any current.

When V_G rises above V_{TN}, transistor M1 turns on and starts to conduct. The current of M1 is determined by (5-24). It is completely intercepted by M3, and the condition of the transistors in the P-subcircuit does not change. However, the potential V_N is starting to decrease.

The trigger operation starts when the voltage V_G arrives at the value of V_{Hi}. At this point, due to simultaneous increase of V_G and decrease of V_N, transistor M2 turns on. It is not difficult to see that if in (5-23) one substitutes V_{dd} in stead of V_{0c} (the gate of M3 is still at V_{DD}) and V_{Hi} instead of V_G, one obtains the required relationship between the transistor parameters to start the triggering operation. It can be rewritten as

$$\frac{k_1}{k_3} = \left(\frac{V_{dd} - V_{Hi}}{V_{Hi} - V_{TNi}}\right)^2 \tag{5-25}$$

By the same reasoning, one obtains that the condition:

$$\frac{k_4}{k_6} = \left(\frac{V_{Li}}{V_{dd} - V_{Li} - |V_{TP}|}\right)^2 \tag{5-26}$$

The voltages V_{Hi} and V_{Li} considered as true thresholds of the CMOS Schmitt trigger. However, in effect, V_{Hi} and V_{Li} only function at the beginning of the triggering operation. The real triggering occurs at close but different voltages V_H and V_L. The difference depends on choice of the parameters k_2 and k_5, and can be estimated as follows.

The transition from one stable state to another in the Schmitt trigger is, indeed, very fast, and one can consider that during it the trigger input voltage does not change and stays at V_H for the considered transition of the output voltage from high to low. When M2 is turned on, the trigger starts to operate as a linear circuit with positive feedback. Transistors M4 and M5 are in a linear mode of operation, and the trigger can be represented as the linear circuit shown in Fig. 5.12(a). The current of M1 is

$$I_{NH} = k_1(V_H - V_{TN})^2 \approx k_1(V_{Hi} - V_{TN})^2 \quad (5\text{-}27)$$

The trigger load is

$$R_{LP} = \frac{k_4^{-1} + k_5^{-1}}{2(V_{dd} - V_{Hi} - |V_{TP}|)} \quad (5\text{-}28)$$

The small-signal model for this part of trigger operation is shown in Fig. 5.12(b). The loop-transfer function for this circuit is

$$A_L = \frac{g_{m3} R_{LP}(g_{m2} r_{01} + 1)}{(g_{m2} + g_{m3}) r_{01} + 1} \quad (5\text{-}29)$$

where r_{01} is the output impedance of M1 that is operating in the saturation region.

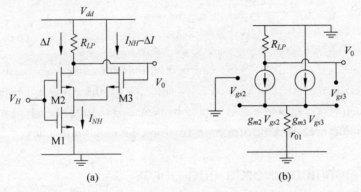

Fig. 5.12 Schmitt trigger during transition: (a) equivalent circuit and (b) small-signal model

At the moment of the output voltage jump from high to low, this loop transfer function becomes equal to unity. Assuming $g_{m2} r_{01} \gg 1$ and there is

$$\frac{R_{LP}}{g_{m2}^{-1} + g_{m3}^{-1} + (r_{01} g_{m2} g_{m3})^{-1}} = 1 \quad (5\text{-}30)$$

The current of M1 at this instant is divided between M2 and M3 into two parts ΔI and $I_{NH} - \Delta I$ so that the transconductance of the corresponding transistors are

$$g_{m2} = 2\sqrt{\Delta I \cdot k_2} \quad (5\text{-}31)$$

$$g_{m3} = 2\sqrt{(I_{NH} - \Delta I) \cdot k_3} \approx 2\sqrt{I_{NH} \cdot k_3} \quad (5\text{-}32)$$

the (5-30) can be simplified as:

$$g_{m2}^{-1} + g_{m3}^{-1} \approx R_{LP} \quad (5\text{-}33)$$

put (5-31) and (5-32) into (5-30), with (5-27) and (5-28) can get:

$$\Delta I = k_2^{-1} \left[\frac{k_4^{-1} + k_5^{-1}}{V_{dd} - V_{Hi} - |V_{TP}|} - \frac{1}{(V_{Hi} - V_{TN})\sqrt{k_1 k_3}} \right]^{-2} \quad (5\text{-}34)$$

ΔI depends on k_2 and k_5. We can estimate the difference between V_{Hi} and V_H. In fact, when the transition starts one has the input voltage of V_H, transistor M2 has zero current, transistor M3 carries the current of I_{NH}, and the trigger output voltage is equal to V_{dd}. Just before the output voltage jump, one has the input voltage of V_H, transistor M2 has the current of ΔI, M3 carries the current of $I_{NH} - \Delta I$, and the output voltage drops to $V_{dd} - \Delta I R_{LP}$. Using these conditions, it is easy to find that

$$\Delta V_H = V_H - V_{Hi} \approx \sqrt{\frac{\Delta I}{k_2}} - \Delta I R_{LP} \tag{5-35}$$

If transistor M3 is very wide we can use the approximation

$$\Delta I \approx \frac{(V_{dd} - V_{Hi} - |V_{TP}|)^2}{k_2(k_4^{-1} + k_5^{-1})^2} \tag{5-36}$$

if $\Delta I R_{LP}$ is neglected, put (5-36) into (5-35), finally we can obtain

$$\Delta V_H \approx \frac{V_{dd} - V_{Hi} - |V_{TP}|}{k_2 k_4^{-1} + k_2 k_5^{-1}} \tag{5-37}$$

Using the same method, considering the transition of the output voltage from low to high, we can get:

$$\Delta V_L \approx V_L - V_{Li} \approx -\frac{V_{Li} - V_{TN}}{k_5 k_2^{-1} + k_5 k_2^{-1}} \tag{5-38}$$

The values given by (5-37) and (5-38) can be considered as the worst case deflections of the thresholds. It is seen that to reduce ΔV_H and ΔV_L, the ratio k_2/k_5 should be kept constant and each of k_2/k_4 and k_5/k_1 should be increased simultaneously. Equations (5-37) and (5-38) provide necessary information for the CMOS Schmitt trigger design.

5.6 Technical words and phrases

5.6.1 Terminology

System-On-Chip(SoC)	片上系统
automatic gain control loops(AGC)	自动增益控制环路
peak detector	峰值检测器
resolution	分辨率
kick-back noise	回踢噪声
input common mode range	输入共模范围
output swing	输出摆幅
hysteresis comparator	迟滞比较器
latch	锁存器
Schmitt trigger	施密特触发器
bistable multivibrator	双稳态多谐振荡器

5.6.2 Note to the text

(1) The main function of comparator circuit is to compare an analog signal with another or reference signal, and output it to get the high and low voltage as binary signal

through comparison processing.

比较器电路的主要功能在于将一个模拟信号与另一个模拟信号或参考信号进行对比，并输出经过比较处理得到高低电平，作为二进制信号输出。

(2) Actually comparators can not identify the tiny voltage difference without limitation. Due to the limitation of finite gain, there is usually a minimum resolvable voltage difference, which is called the accuracy(or resolution) of comparators.

在实际电路中，比较器并不能无限制地分辨微小的电压差别，由于有限增益的限制，往往存在一个最小的可分辨电压差，这称为比较器的精度(或分辨率)。

(3) The delay is generally defined as the time difference between the input analog signal and the output digital signal. This parameter determines the maximum operating frequency of comparator.

传输延迟时间一般定义为输入模拟信号与输出数字信号之间的时间差。该参数决定了比较器的最高工作频率。

(4) It can be seen that the relationship between the resolution and the gain is very close, and the high resolution and high precision comparator circuit also means that its gain is higher.

由此可见，比较器的分辨率和增益之间的关系非常紧密，高分辨率高精度比较器电路也意味着其增益较高。

(5) A high-gain OPA in open-loop state is a high-resolution comparator, and the hysteresis comparator and latch circuit is closed-loop amplifier with positive feedback.

一个高增益的运算放大器工作于开环状态就是一个高分辨率的比较器，而迟滞比较器和锁存器电路则是带有正反馈的闭环放大器。

(6) The open-loop comparator is realized by open-loop amplifier. Such comparators do not need frequency compensation, so that the maximum bandwidth can be obtained. Meanwhile in theory, a relatively fast response time can be acquired.

开环比较器的特点是以放大器的开环应用作为基本比较器电路，这类比较器不需要频率补偿，从而可以获得尽可能大的带宽。理论上也就可以获得相对比较快的输出响应时间。

(7) This double-threshold action is called hysteresis, indicating the memory of Schmidt trigger. In essence, Schmitt trigger can be considered a bistable multivibrator.

这种双阈值动作被称为迟滞现象，表明施密特触发器有记忆性。从本质上来说，施密特触发器也可以认为是一种双稳态多谐振荡器。

Chapter 6 Analog-to-Digital Converter

Since 1990s, digital circuits have gradually replaced analog circuits in many fields, which have become an important part of integrated circuits. However, as the most classic circuit design form, analog circuit still has an unshaken position in many aspects. The Analog-to-Digital Converter (ADC) is one of the typical representatives. ADCs are usually used for sampling and digitalization of sensor signals. The purpose is to enable different current or voltage signals to be processed in the digital domain.

6.1 Summary

ADC is a circuit unit that converts analog signals into digital signals. Its main function is to convert a continuous analog signal in time and amplitude into a digital signal that is discrete in time and amplitude. As shown in Fig. 6.1, the transfer process first completes the discretization of analog signal in time domain by sampling on the uniform time point, and then sends it to the quantizer and encoder to realize the digitalization.

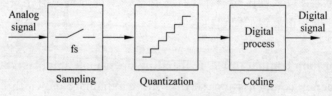

Fig. 6.1   ADC basis

Since the quantizer is a nonlinear system, quantizer generates quantization errors in the quantization process and influences the performance of ADC by quantization noise. Fig. 6.2 is an ideal 3-bit ADC transfer curve and a schematic diagram of quantization error. As seen from the graph, the quantization error is within the range of $\pm \Delta/2$, and Δ is the quantization step.

The input full range of ADC is Full Scale(FS) and the resolution is N, then the input full range is divided into 2^N quantized steps. So the quantization step is as follows:

$$\Delta = \frac{FS}{2^N} \tag{6-1}$$

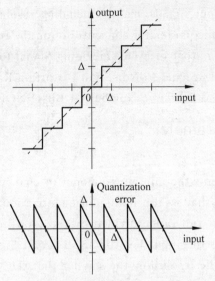

Fig. 6.2 Transfer characteristics and quantization
error of ideal ADC

The input analog voltage is V_a and the output digital signal is D_{out}, then the transfer function of ADC can be expressed as:

$$\frac{V_a}{\Delta} = D_{out} + q_e = \sum_{m=0}^{n-1} B_m 2^m + q_e \quad (6-2)$$

where B_m is the output digital code, q_e is quantization error.

When the ADC resolution is high, it can be assumed that the quantization error is approximately white noise. And the same distribution probability is presented in the range, so quantization error can be called quantization noise, and its power is

$$P_n(t) = \int_{-\infty}^{\infty} e^2 \cdot p(e,t) \mathrm{d}e \quad (6-3)$$

in (6-3), e is quantization error, $p(e, t)$ is the probability density function of quantization error. Assuming that the quantization noise is uniformly distributed in the range of $\pm \Delta/2$ and does not change with time, it can be obtained:

$$p(e,t) = \begin{cases} \dfrac{1}{\Delta} & -\dfrac{\Delta}{2} < e < \dfrac{\Delta}{2} \\ 0 & \text{others} \end{cases} \quad (6-4)$$

Noise power P_n is:

$$P_n = \int_{-\Delta/2}^{\Delta/2} e^2 \cdot \frac{1}{\Delta} \mathrm{d}e = \frac{1}{\Delta}\left(\frac{e^3}{3}\bigg|_{-\Delta/2}^{\Delta/2}\right) = \frac{\Delta^2}{12} \quad (6-5)$$

6.2 Performance parameters

The parameters that measure the performance of ADC can be divided into static and dynamic parameters. Static parameters are focused on measuring the accuracy of ADC for the quantization of input signals, mainly including resolution, linearity, offset, gain error and so on. The dynamic parameter concentrates on the relationship between signal, harmonic and noise in the output of ADC, including Signal-to-Noise Ratio(SNR), Total Harmonic Distortion(THD), Signal-to-Noise and Distortion Ratio(SNDR), Spurious Free Dynamic Range(SFDR) and Effective Number Of Bits(ENOB).

6.2.1 Static parameter

1. Resolution

The resolution represents the minimum amount of change in the analog input signal that ADC can distinguish, that is, the minimum quantization capability. The resolution can be defined as the input full range divided by the total interval number, that is, the minimum distance between the input level is 1LSB(Least Significant Bit). If the input range is fixed, the higher the resolution, the smaller the ADC distinguishable signal, and the higher the resolution.

2. Differential Nonlinearity(DNL)

DNL is the deviation between the actual quantization step of two adjacent numbers and the ideal quantization step(1LSB). Its mathematical expression is as follows:

$$DNL(D_i) = \frac{V_{in}(D_i) - V_{in}(D_{i-1}) - V_{LSB}}{V_{LSB}} \quad (6\text{-}6)$$

D_i are D_{i-1} two adjacent digital output codes, and $V_{in}(D_i)$ and $V_{in}(D_{i-1})$ are the analog input voltages for these two digital codes. V_{LSB} is the ideal quantization step $V_{LSB} = V_{FS}/M = V_{FS}/2^N$. Fig.6.3 is a schematic diagram of DNL in which the input signal amplitude is 2V and the resolution is 3 bits. For an ideal ADC, the interval between two adjacent voltage is 1LSB, that is, the DNL is 0LSB. In Fig.6.3 in the range from 1000 to 1500mV of the input signal, the actual quantization step is 2.2LSB, that is, DNL is 1.2LSB here. It can be found that no matter the input of any analog signal, this ADC can not output digital code of 100, which is the phenomenon of missing code. Generally, when $|DNL| \geq 1LSB$, missing code would happen. It is necessary to point out that DNL is related to SNR, but SNR can not be predicted by DNL.

3. Integral Nonlinearity(INL)

DNL is defined as the deviation between the actual tranfer characteristic curve and the ideal transfer characteristic curve. It indicates the maximum degree of deviating from the ideal curve of the actual transfer curve. Fig.6.4 is the INL diagram in which the input signal amplitude is 2V and the resolution is 3 bits. The maximum difference between the two transfer curves identified in the graph is 0.6LSB, so the maximum INL is 0.6LSB.

INL is related to the total harmonic distortion, which can most reflect the linearity of ADC. For a 10 bit ADC if its INL = 4LSB, it indicates that the ADC has a 8 bit linearity. It means that ADC can only be used as a 8 bit ADC.

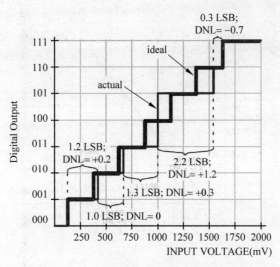

Fig. 6.3 DNL

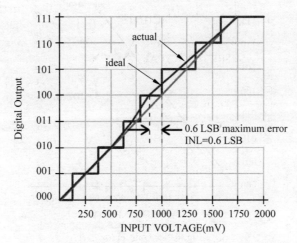

Fig. 6.4 INL

4. Offset

The offset voltage and current brought by input amplifier, output amplifier and comparator are inevitable, while the ADC are made up of the above unit modules. Therefore, offset is also one of the parameters that need to be considered in the application of data conversion system, especially in the DC application system. As shown in Fig. 6.5, the definition of offset is that when the input signal reaches a certain value, the output changes from the smallest digital code. At this time the input signal voltage is offset.

5. Gain error

Gain error is demonstrated in Fig. 6.6. In order to eliminate the offset influence on gain error measurement, the starting point of the transfer curve was shifted to original point. So the starting point of the transfer curve is the same between the ideal and the actual, and the difference between the slope of the two curve is gain error.

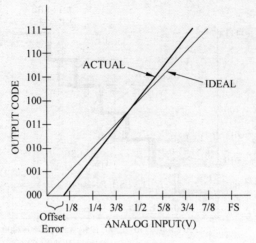

Fig. 6.5   Offset

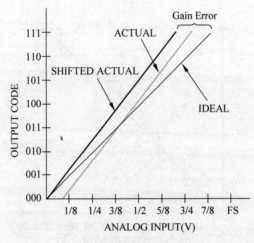

Fig. 6.6   Gain error

6.2.2 Dynamic parameter

The dynamic characteristics of ADC are related to its conversion rate, the frequency and amplitude of input signal, which are usually measured in the frequency domain. The dynamic characteristic parameter definition of ADC is shown in Fig. 6.7.

1. Signal-to-Noise Ratio(SNR)

In a certain frequency SNR is the power ratio of output signal and noise:

$$\text{SNR} = 10 \cdot \log \frac{P_s}{P_n} \tag{6-7}$$

P_s and P_n respectively denote the signal power and total noise power. The total noise power includes both quantization noise and other noise introduced in analog circuit. The unit of SNR is decibels.

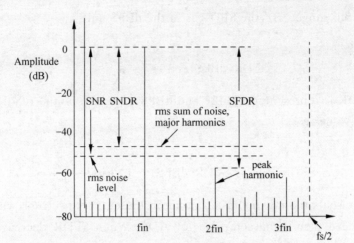

Fig. 6.7 The dynamic characteristic parameter definition of ADC

The sine signal is often used to measure ADC. The power of a single frequency sinusoidal signal with amplitude of $\Delta \cdot 2^{N-1}$ is:

$$P_s = \frac{(\Delta \cdot 2^{N-1})^2}{2} \tag{6-8}$$

When the sinusoidal signal is processed by an ideal ADC, SNR is:

$$\mathrm{SNR} = \frac{P_s}{P_n} = \frac{(\Delta \cdot 2^{N-1})^2 / 2}{\Delta^2 / 12} = 1.5 \cdot 2^{2N} \tag{6-9}$$

The conversion to decibels(dB) is as follows:

$$\mathrm{SNR} = 10 \cdot \log \frac{P_s}{P_n} = 6.02 \cdot N + 1.76 \mathrm{dB} \tag{6-10}$$

In (6-10), it can be seen that when the resolution of ADC increases by 1 bits, the SNR will be increased by about 6dB.

2. Signal-to-Noise-and-Distortion Ratio(SNDR)

SNDR is the ratio of output signal power to harmonic and total noise power in a certain frequency band.

$$\mathrm{SNDR} = 10 \cdot \log \frac{P_{signal}}{P_{Noise} + P_{Distortion}} \tag{6-11}$$

SNDR represents the performance degradation of ADC due to all kinds of noise and harmonic distortion. SNDR is usually corresponding to a sinusoidal signal, which is related to the frequency and amplitude of input signal.

3. Spurious-Free Dynamic Range(SFDR)

SFDR is the ratio of the signal power to the maximum harmonic(in a unit of dBc)

$$\mathrm{SFDR(dBc)} = 10 \cdot \log \frac{X_{sig}^2}{X_{spur}^2} \tag{6-12}$$

X_{sig}、X_{spur} are the root mean square(RMS) of the fundamental frequency signal and the maximum harmonics, respectively. SFDR is related to the signal amplitude. When the input signal is full range(FS), the SFDR is in the dBFS unit.

$$\text{SFDR(dBFS)} = 10 \cdot \lg \frac{\left(\frac{\text{FS}}{2\sqrt{2}}\right)^2}{X_{spur}^2} \quad (6\text{-}13)$$

When SFDR is represented by dBc and dBFS respectively, the relationship between the two SFDR values is:

$$\text{SFDR(dBc)} = \text{SFDR(dBFS)} - 10 \cdot \log \frac{\left(\frac{\text{FS}}{2\sqrt{2}}\right)^2}{X_{sig}^2} \quad (6\text{-}14)$$

SFDR represents the maximum harmonic in the output spectrum, which has a great influence on the dynamic characteristics of ADC. When SFDR decreased to a certain extent, it will make other dynamic characteristics significantly decreased.

4. Total Harmonic Distortion(THD)

THD is expressed as the ratio of the total harmonic distortion power to the fundamental composition(output signal) power.

$$\text{THD} = 10 \cdot \log \frac{P_{THD}}{P_{Signal}} = 10 \cdot \log \left(\sum_{k=2}^{\infty} X_k^2 / X_1^2 \right) \quad (6\text{-}15)$$

X_1 is the root mean square of the fundamental composition, and X_k is the root mean square of the K-harmonic. The first 10 harmonics are usually used in the calculation of THD. The THD reflects the deterioration of SNR caused by the harmonic distortion.

5. Effective Number Of Bit(ENOB)

ENOB is calculated by SNDR of the full range input signal, and the mathematical expression is as follows:

$$\text{ENOB} = \frac{\text{SNDR} - 1.76}{6.02} \quad (6\text{-}16)$$

6.3 Flash ADC

The flash ADC is a fully parallel structure, which is the fastest ADC structure at present. Fig. 6.8 is a structure block diagram of a N-bit flash ADC. The circuit consists of $2^N - 1$ comparators, 2^N voltage dividers, and a decoder. The voltage divider divides the voltage reference into the N section, and the comparator compares the input voltage V_{in} with these voltages at the same time. Each comparator produces digital output code, and the code output by all comparators is a thermometer code. The thermometer code is converted through the decoder into binary code or gray code.

The advantages of flash ADC are fast conversion and simple structure. While its disadvantage of simple structure is that the number of comparators is exponentially related to resolution, so its power consumption, area and input capacitor are exponentially related to resolution. For example, a 10 bit flash ADC has 1023 comparators and 1024 resistors, which is a great challenge to area and power consumption. At the same time, the resolution of the comparator is difficult to control because of the offset and disturbance of voltage reference. Due to the limits of area, power, and resolution, flash ADC is usually suitable for 4~8 bit ADC design.

To eliminate the limitations on the broadband application systems, in recent years, the revolutionary development of a variety of factors has promoted the continuous innovation of medium resolution Flash ADC architecture and circuit. Many hybrid ADC solved the technical problems in the traditional structure, which made a great progress for medium resolution Flash ADC.

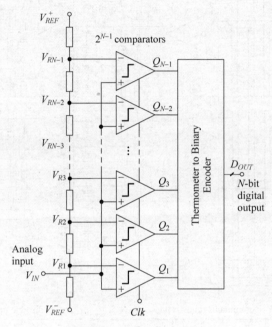

Fig. 6.8 N-bit flash ADC

6.3.1 Flash-interpolation structure

The traditional flash ADC is a fully parallel structure, which requires $2^N - 1$ preamplifiers / comparators to realize N-bit resolution, and the large input capacitor seriously limits the input signal bandwidth. The interpolation technique can reduce the number of preamplifier, reducing the input capacitor and optimizing the input signal bandwidth. The interpolation technique based on resistor network is to connect two resistors in series between two amplifiers output terminals to generate zero due to the decrease of amplifiers, so as to achieve N-bit data conversion. The resistor network used to implement the interpolation technique can be used as a resistor network in the offset averaging technique. Therefore, the flash-interpolation is the most commonly used structure to realize GSPS flash ADC with wide signal bandwidth.

A typical analog front-end circuit for a flash-interpolation ADC is shown in Fig. 6.9.

The circuit includes a cascade of four-stage preamplifier array. The output of the first three amplifier uses 2-times resistance interpolation, and the fourth output realizes the 6-bit resolution. The number of voltages reference is reduced from 65 to 11 by interpolation. In flash-interpolation structure, the offset averaging technique has always been the focus of research.

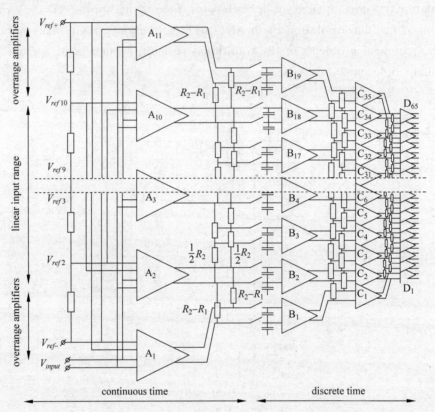

Fig. 6.9　A typical analog front-end circuit for a flash-interpolation ADC

The flash-interpolation structure using the resistor interpolation technique adopts a multistage preamplifier array, which consumes more power, and the energy efficiency is very low. However, because the structure has the characteristics of high-speed sampling, interpolation technique is used to optimize the parasitic effect of nodes on the signal path, which will fully reflect the speed advantage of this structure.

6.3.2　Folding-interpolation structure

The interpolation technique can reduce the number of preamplifiers and the input capacitor, but the number of comparators is the same as the flash structure. while folding technique can reduce the number of comparators. The folding and interpolation structure combines two technical advantages, which reduce the input capacitor, chip area and power consumption, and the dynamic performance is also improved.

The typical folding and interpolation structure, as shown in Fig. 6.10, uses resistor interpolation in the output of the folding amplifier, and the interpolation coefficient is 4. The number of zero required for 2-bit resolution is realized at the fold stage output. After interpolation, the number of zero corresponding to the 6-bit resolution is realized. The Folding-interpolation structure can effectively improve the resolution and speed by reducing the influence of non ideal factors by using the calibration technique.

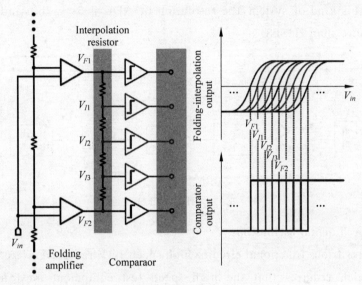

Fig. 6.10 Folding-interpolation structure

6.3.3 The main application areas of Flash ADC

As modern information throughput is increasing, these information processing devices need wider signal bandwidth to transmit high throughput data. In these devices, the medium resolution and GSPS flash ADC are an essential module. In the application fields of data storage system, high-speed digital oscilloscope, broadband communication system, radar, medical imaging, high-speed data acquisition system, the performance requirements for medium resolution and GSPS flash ADC are very high.

1. Data storage system

In general, various data storage systems use magnetic or optical signals to read and write data through the read channel and write channel. The Hard Disk Drive(HDD) and Digital Versatile Drive(DVD) are typical data storage systems. Fig. 6.11 shows a schematic diagram of a HDD read channel. The "read signal" is an electrical signal converted from magnetic signal information(HDD) or optical signal information(DVD). The variable gain amplifier(VGA) is used to adjust the "read signal" amplitude to satisfy ADC full input swing. Low-Pass Filter(LPF) is used to attenuate high frequency signal to remove aliasing effect. The Viterbi detector compares the sampled digital codes output by ADC with all possible bit sequences to achieve accurate signal detection. In read channel circuits, ADC and FIR equalizers consume a large amount of area and power. The scaling of CMOS device can make the FIR equalizer reduce power and area effectively, but it brings great challenge to the ADC design. As the data processing speed and data storage capacity of the storage system are increasing, the sampling frequency of ADC is also increasing. In this kind of system, the resolution of ADC is 5 ~ 7bit, and the sampling frequency is more than 1GSPS.

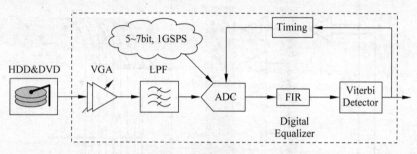

Fig. 6.11　HDD read channel

2. High-speed digital oscilloscope

At present, various functional circuits(high speed I/O transceivers, etc.) work faster and faster, which requires that the high speed test equipment must have a higher operating frequency. High-speed digital oscilloscope is a typical application example, as shown in Fig.6.12.

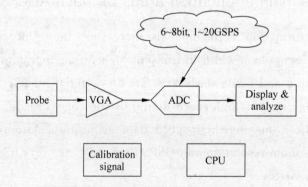

Fig. 6.12　High-speed digital oscilloscope

The probe is used to detect the external signal, which is amplified by a preamplifier, and then digitized by ADC. The calibration signal is supplied externally, and the calibration coefficient is calculated by CPU inside the system to correct the nonlinear characteristics of ADC. The time interleaving technique is used to improve the sampling frequency of GSPS flash ADC for high-speed digital oscilloscope. Generally, the resolution of ADC is more than 6bit, and the sampling frequency is 1~20GSPS.

6.4 Two-step ADC

Due to the exponential growth of power consumption, area and input capacitor with the resolution, the structure of flash ADC is not suitable for achieving more than 8 bits resolution. In order to apply flash structure to high resolution ADC design, a two-step structure ADC is developed.

As shown in Fig. 6.13, the two-step ADC is composed of a sample and hold amplifier (SHA), two low-resolution flash ADC, a DAC, a subtracting circuit and a residual amplifier. SHA samples the input signal during the sampling phase; At the holding phase, the sampling signal is converted to a high digital code B_1 by the coarse flash ADC. The digital code is sent to a DAC to be converted to an analog signal and subtracted from the original input signal, then the result of the subtraction is called the residual. After that, the residual is multiplied by 2^{B_1} and then input to the fine flash ADC to get the low digital code B_2. The function of residual amplification is to make the residual voltage reach the same swing as the original signal, so that the fine process can use the low-resolution flash ADC. Such a structure simplifies the circuit design.

The greatest advantage of the two-step ADC structure is that the number of comparators is greatly reduced compared with flash ADC. If the coarse ADC and the fine ADC provide the same number of digital code, the two-step ADC of a N-bit requires only $2(2^{N/2} - 1)$ comparators. Since the number of comparators increases with the resolution, the two-step structure can be used on the 10 bit resolution ADC. The drawback of the two-step structure is that the two-step conversion requires two clock cycles to complete, which limits the speed of this structure.

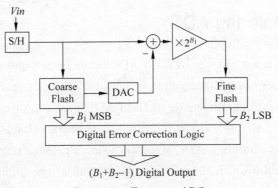

Fig. 6.13 Two-step ADC

6.5 Folding ADC

The principle of folding ADC is to use preprocessing technique to reduce hardware consumption and to maintain one-step conversion characteristic of flash ADC. Similar to the two-step ADC, the folding ADC needs coarse and fine ADC. The analog signal is converted to the high bit through a coarse ADC. Meanwhile, another fine ADC is used to convert the folded signal to the low bit. Unlike the two-step structure, the two ADC works at the same time. It only takes a clock cycle to complete the conversion, so the conversion speed is fast.

Fig. 6.14 is a folding ADC with a factor of 8. From the ideal folding waveform, it can be seen that the range of the signal is reduced to 1/8 of original, and the number of required comparators is also changed into 1/8. Although the folding structure uses a clock cycle less than the two-step structure, its speed increase is not two times. This is due to folding once, and the frequency of the folding output is doubled. The bandwidth of the folding circuit must be twice as large as the input signal, which is limited by the bandwidth of the folding circuit. In addition to this, the folding circuit will also bring about a nonlinear problem. The timing error between the folding amplifier and the coarse ADC can cause the deviation of the folding curve, which will affect its performance.

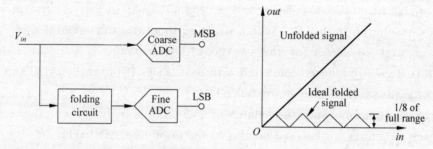

Fig. 6.14　Folding ADC priciple

6.6　Interpolating ADC

One of the main disadvantages of flash ADC is the large input capacitance, while the interpolation technique can effectively reduce the input capacitance. Interpolation technique uses the linear characteristics of flash ADC near the threshold voltage, through resistor series, current mirrors, capacitors to generate more voltage reference.

Fig. 6.15 is an interpolating ADC structure using a resistor string as an interpolating network. The interpolation technique greatly reduces the number of comparators connected to the input, thus reducing the input capacitance. Besides the interpolating ADC does not need the sampling circuit inside, and can continue to maintain the one-step conversion characteristic of flash ADC with fast operating speed. But the interpolating structure alone can not reduce the number of latches. Therefore, interpolating ADC is usually combined with folding technique, so as to reduce the number of pre-amplifiers and latches at the same time, and further reduce the area and power consumption.

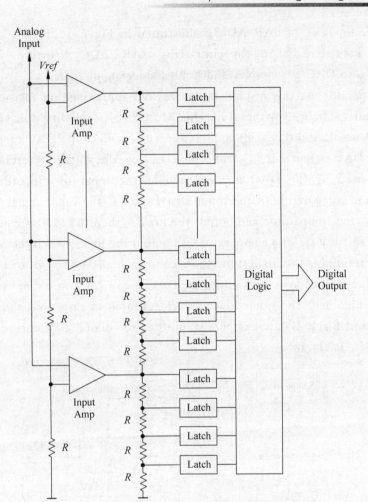

Fig. 6.15 Interpolating ADC structure using a resistor string as an interpolating network

6.7 Successive Approximation Analog-to-digital Converter(SAR ADC)

SAR ADC is an analog-to-digital converter with medium sampling rate(about 1~50 MSPS) and medium resolution(10-18 bit). Because of its simple structure and low power consumption, it has been widely used in the field of sensor detection and industrial control.

The principle of SAR ADC is the application of binary search algorithm, that is, the two-difference algorithm is used to approach the analog input step by step. SAR ADC is mainly composed of timing generator, sample and hold circuit, Digital-to-Analog Converter(DAC), comparator and Successive Approximation Register(SAR). DAC and comparator are the two most important modules, which determine the resolution and speed.

The basic topology of SAR ADC is illustrated in Fig. 6.16. In this structure, first, sample and hold circuit (in capacitor-array SAR ADC structure, which can be incorporated into DAC's), samples and holds analog input signal V_{IN}, and use it as an input of comparator. At this instant, the successive approximation register (SAR logic) starts the binary search algorithm. First, the Most Significant Bit (MSB) is set to 1, and the other bits are 0. And the N bit code $(100,\cdots,0)$ is add to the DAC capacitor array, at this time the DAC output $1/2 V_{REF}$, of which V_{REF} is the voltage reference. The analog voltage converted by the DAC is then used as the input of the other end of the comparator and compared with the input signal V_{IN}. If the input signal V_{IN} is greater than $1/2 V_{REF}$, the comparator will output the logic 0, then the MSB remains 1; While if V_{IN} is less than $1/2 V_{REF}$, the comparator will output the logic 1, then the MSB will be set to 0. After determining the highest bit, the secondary highest bit is set to 1, the other low bits are 0. Then the code is also added to DAC array for comparing and the second highest bit is obtained. The other low bits are repeated in turn until the results of the Least Significant Bit (LSB) are compared, and then the digital code corresponding to the input signal V_{IN} is obtained.

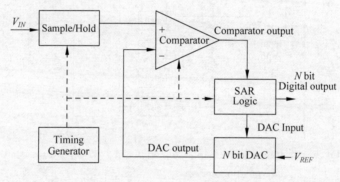

Fig. 6.16 SAR ADC

A typical charge-redistribution DAC of 10bit is shown in Fig. 6.17. The segmented capacitor C_d is a unit capacitor 1C, and the sampling capacitor CS is 64C, and the total equivalent capacitor is 128C. The DAC is sampled by a pulse signal CLK1 that lasts two clock cycles, and the operating process is as follows: the operating process of DAC is divided into sampling and charge-redistribution phase two stages. In the sampling phase, the S0~S9 switch are connected with GND. At the same time switch sample is closed, so that the lower plate of capacitor CS is connected with the VIN. While Svcm is closed, the upper plate is connected with the common-mode voltage VCM. Then the charge stores on the sampling capacitors CS. After the sampling phase and in the holding phase, the switch Svcm is off, the switch Ssample and switch S0~S9 are connected with GND. So the DAC output voltage for the time is:

$$V_x = \frac{Q_x}{C_t} = \frac{64C(v_{cm}-v_{in})+v_{cm}(15C /\!/ 1C + 63C)}{15C /\!/ 1C + 127C} = -\frac{1024}{2047}v_{in} + v_{cm} \qquad (6\text{-}17)$$

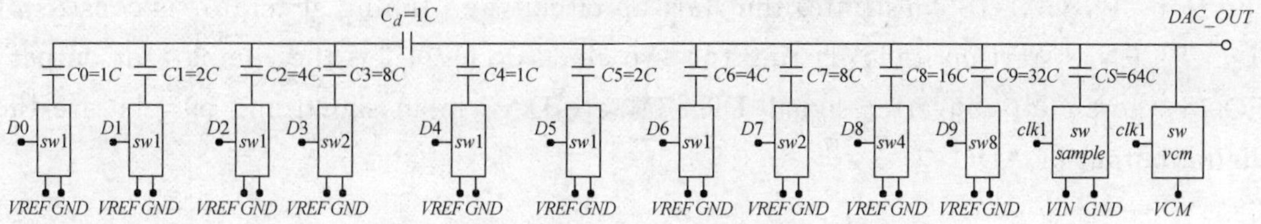

Fig. 6.17 DAC

In the charge-redistribution phase, firstly the tenth bit (MSB) is set to 1, which means lower plate of C9 is connected to VREF through the switch S9. If $v_{in} > 1/2 * v_{ref}$, the output of the comparator is 0 and the MSB remains 1; Otherwise the MSB is set to 0. After that the ninth bit is set to 1 and it repeats operation. Then the eighth bit, the seventh bit…. until the first bit (LSB) is determined. Finally the output of DAC is:

$$V_x = \frac{1024}{2047}\left(-v_{in} + \sum_{i=1}^{10} \frac{b_i}{2^{11-i}} v_{ref}\right) + v_{cm} \tag{6-18}$$

b_i is the value of the i bit, which is 0 or 1.

Although the offset of comparator will not affect the linearity of ADC conversion, it will still cause ADC's missing code. Therefore, some offset reduction techniques are adopted in design. A comparator with input offset storage and output offset storage is shown in Fig.6.18, which mainly includes three pre-amplifier, latch, and input and output offset storage capacitors.

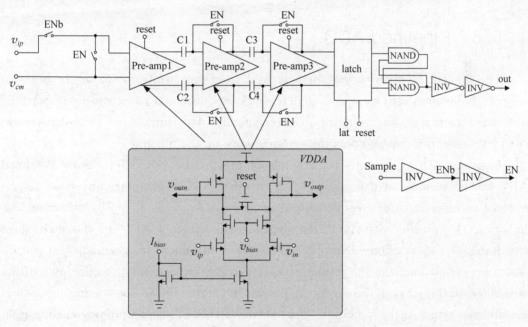

Fig. 6.18 Comparator

SAR logic circuit is the most important part of SAR ADC. It realizes the feedback control of DAC. The SAR logic circuit comprises D flip-flops, inverters, AND gates and JK flip flops, as shown in Fig.6.19. F0~F9 are 10-bit SAR logic registers made up of JK flip-flops. FS, GA, GB constitutes the start-up circuit, and timing generator is consists of FA~FJ. EN is start-up signal lasting for two clock cycles. V_c is the comparator output. EOC is the end-of-converter signal. D9~D0 are DAC input signal and b9~b0 are the digital output of ADC.

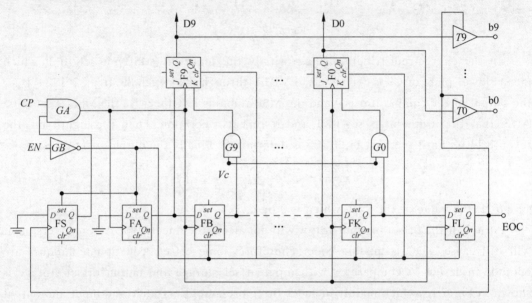

Fig. 6.19 SAR logic circuit

6.8 Pipelined ADC

The concept of pipeline was first derived from industrial production. In order to improve production efficiency, a complete work was decomposed according to working order, and each part was assigned to corresponding departments. So each department completes a specific job and puts the finished work on the pipeline.

Pipelined ADC is a successful example of pipeline in ADC applications. Pipelined ADC can be seen as an extension of the two-step ADC. The total conversion rate is reduced because a complete conversion of two-step ADC is completed by two sub-ADCs in series. The pipeline structure first expands the two-step ADC to the multi-stage structure. After that, a sample and hold amplifier is introduced between adjacent stages, which keeps and amplifies the residue after last conversion, and serves as the input of the next stage. In this way, all the stages can process in parallel the analog residue obtained at the previous stage. At the same time, due to the existence of gain amplifying circuits at all stages, the requirements for the post stage resolution can also be reduced. From the perspective of conversion process, all stages of pipelined ADC use serial processing mode, each stage's input is the output of last stage. Only after one stage work is completed, the next stage can start work. But in every step conversion, every stage is working at all the time, no one is resting, so the working mode of each stage can be regarded as parallel.

The working mode of the pipelined ADC determines that it has two features: first, the pipelined ADC decomposes the conversion into a few stages. In this way, when the analog input is entered into the pipelined ADC, a number of clock cycles will be waiting to get the digital output. There is a delay, the delay is proportional to the number of stages; secondly, the maximum operating frequency is mainly limited by the speed of each stage. If there is a difference at all stages, the slowest stage deteimines the fastest speed of the pipeline ADC.

The structure of a typical pipelined ADC is shown in Fig. 6.20. As can be seen from the figure, the pipelined ADC is composed of k low-resolution stages, delay element and digital correction circuit. Stage i outputs $B_i + r_i$ bit digital code, where B_i is the effective conversion bit and r_i is redundant bit. Redundant bits are introduced to solve the non-ideal factors in the circuit, especially the error caused by the comparator offset. Each stage's B_i can select the same or different values. In combination with resolution of each stage, B_i selection for each stage is reduced step by step. The redundant bits r_i can be chosen based on the tolerance of the comparator. When $r_i = 1$, the tolerance of pipelined ADC to the comparator offset reaches the maximum.

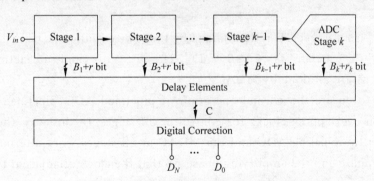

Fig. 6.20 The structure of pipelined ADC

The last stage of pipelined ADC can have two options: The first one is to select the stage with redundant bits, which requires that if this stage can provide the effective bits, the redundant bits can be abandoned directly because they are not used. For example, a pipelined ADC of 12bit requires the last stage to provide two effective bits. At this time, if 2.5bit's stage structure is adopted and two effective bits are obtained, the redundant bits can be abandoned directly; Another option is to use the last stage without redundancy, such as flash ADC. This is because the residue generated at the last stage is no longer useful, no redundant design needs to be made. This choice is relatively simple and has a smaller power consumption, so it is widely used. In the last stage of high-resolution pipelined ADC, the multi-output flash structure is used, and the output bits are selected as 2~4 bit.

The structure of each stage in pipelined ADC is shown in Fig. 6.21. It is made up of a low-resolution ADC and a Multiplying DAC(MDAC). The low-resolution ADC usually uses a flash ADC structure.

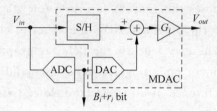

Fig. 6.21   The structure of each stage in pipelined ADC

Its functions include sampling and holding, digital-to-analog conversion, subtraction operation and amplification. In the data conversion process, each stage first performs analog-to-digital conversion of analog signals and generates $B_i + r_i$ bits output. Meanwhile it converts the digital code to analog signal through DAC, and subtracts the input analog signal to get the residue. The residue is further amplified by G_i times and sent to the next stage for the same processing. G_i can be expressed as

$$G_i = 2^{B_i + 1 - r} \tag{6-19}$$

The MDAC transfer function of an ideal $B_i + r_i$ bits output can be expressed as

$$V_{out,i} = G_i \cdot V_{in,i} - D_i \cdot V_{ref} \tag{6-20}$$

where D_i is an integer. It is determined by the low-resolution ADC in each stage and its range is $D_i \in [-(2B_i - 1), +(2B_i - 1)]$. The G_i, $V_{in,i}$, and subtraction operations in the equations are all done by MDAC.

In order to ensure the normal operation of pipelined ADC, pipelined ADC selects two-phase non-overlapping clocks for control at all stages. In this way, the sample and hold circuit in the pipeline ADC and the MDAC at all stages work alternately between the sampling phase and the amplifying phase. So that the clock of adjacent two stages are opposite, it also means the working states of the adjacent two stages are always different. For example, the first stage is in the sampling phase, the later stage is in the amplification phase; while the current stage is in the amplification phase, the latter stage is in the sampling phase.

The non-overlapping clock is generated by a clock generation circuit. It is assumed that the non-overlapping clock is Φ_1 and Φ_2. The Φ_1 controls sample and hold circuit and the MDAC of even number, and the Φ_2 controls the MDAC of odd number. When the clock Φ_1 is high, the sample and hold circuit and the MDAC of even number are in the sampling phase, while the MDACs of odd number are in the amplifying phase. At this time the sample and hold circuit samples the input analog signals, while the even MDAC samples the analog residual of the odd MDAC amplification output; When the clock Φ_2 is high, the sample and hold circuit and the even MDACs are in the amplification phase, while the odd MDACs are in the sampling phase. The above process is operated repeatedly, and the input signal is processed by the MDACs of different stage. Because the digital output of each MDAC is not synchronized, it is necessary to use the delay element to synchronize. In the structure of redundant bits, we need to reconstruct the digital codes obtained at each stage, which is the function of the digital correction module. While in a structure that does not use redundant bits, the synchronized digital code can be output directly.

6.8.1 Sample and hold

In pipelined ADC, sub ADC and MDAC are used to complete the function of the analog-to-digital conversion. In theory, no other modules are needed. As the output of each stage MDAC has the function of holding, the performance is stable from the second stage. and the problem may exist before the first MDAC.

As shown in Fig. 6.21, analog signal is divided into two parts after entering the first stage, the first is to enter the sub ADC to generate the digital code, and then converted by DAC for subtraction; the other goes directly into the MDAC. Better performance can be achieved if the two signals are well matched. If there is an error between them, it will result in the decline of performance and even the realization of function. Suppose that a sine signal with an amplitude of A is input, its bandwidth is f_{BW} and the maximum offset allowed by MDAC is $V_{tolerance}$. The sum of the offset voltage of the comparator and the inter-stage amplifier is V_{offset}, and the clock timing difference of the two signals is Δt. In order to meet the design requirements for the linearity and dynamic characteristics, the following relationship needs to be satified:

$$f_{BW} \leqslant \frac{V_{tolerance} - V_{offset}}{2\pi \cdot A \cdot \Delta t} \quad (6-21)$$

From (6-21), it is known that if the bandwidth of the signal is larger, the tolerance of the offset is greater and the clock timing difference of the two signals is smaller. These two points need to be solved in structure selection and timing implementation. However, it is difficult to meet the above two points in the case of fast input signal. So the current solution is to increase the sample and hold circuit in front of the first stage. In this way, in the first stage of ampling phase, the analog signals that are input to the two channels can be kept constant. Even if the clock of the two signals is quite different, it will not affect the performance of the first stage.

Fig. 6.22 is a sample and hold circuit by using a simple sampling switch, which is the simplest sample and hold structure. When the switch is on, the input signal charges the sampling capacitor. When the switch is open, the charge stored on the capacitor remains unchanged, so that the output V_{out} is equal to the input V_{in} at the time when the switch is off. Of course, such a hypothesis is obtained when the on-resistance of the switch is zero. In practice, the on-resistance has a certain effect on the accuracy of the sampled signal.

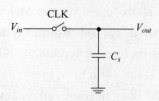

Fig. 6. 22   Sample and hold circuit by using a simple sampling switch

In addition to the influence of the on-resistance on the sampling accuracy, the structure also has a more serious clock feedthrough and charge injection effect. Under the condition of high sampling accuracy, the error introduced by these two effects will seriously affect the sampling accuracy. To eliminate these effects, the bottom-plate sampling is usually used to solve them. Fig. 6.23 is a schematic of the bottom-plate sampling. The bottom plate of the sampling capacitor is connected to the ground by switch. The switch M2A and M2B turn on slightly before M1A and M1B, and the output end is suspended so that the clock feedthrough and charge injection caused by M1A and M1B will not affect the output. Though clock feedthrough and charge injection of switch M2A and M2B will have an impact on output, the error introduced in theory can be eliminated by differential structure. Even though there are differences in the influence of switch M2A and M2B's clock feedthrough and charge injection on the output, the error of these two switches can also be controlled within the resolution requirement range because the size of these switches is generally small.

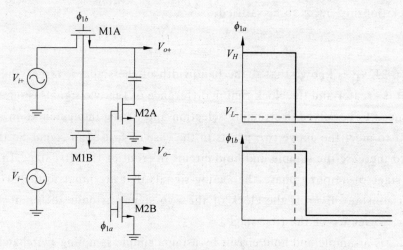

Fig. 6. 23   The bottom-plate sampling and timing

The simple switched-capacitor sample and hold circuit can meet the requirements of general resolution after using the bottom-plate sampling technique. However, for pipelined ADC, this structure can not be used. This is because the charge is stored in sampling capacitor C_S in holding phase by sample and hold circuit. At this point, the MDAC in the first stage needs to execute sampling operation. Because the sampling capacitor of the first stage MDAC will also be connected to the output of sample and hold circuit, which will cause the charge sharing phenomenon. Therefore, this passive switched-capacitor sample and hold structure can not be applied to the pipelined ADC. Current pipelined ADC sample and hold circuits usually are active structures. The characteristic of active sample and hold circuit is to provide continuous current supply to the output terminal in holding phase, so as to ensure charging for the sampling capacitor of the first MDAC.

6.8.2 Sub-ADC

In a pipelined structure, the $B_i + r_i$ bit quantization at each stage is done by a sub-ADC. The sub-ADC generally uses a flash structure as shown in Fig. 6.24. The comparison voltage is obtained by the resistor divider. The output of each comparator is the thermometer code, and the binary code can be obtained after the encoder.

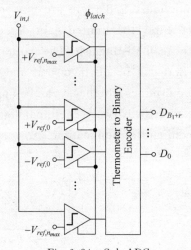

Fig. 6.24 Sub-ADC

In order to use the bottom-plate sampling technique to eliminate the charge injection and clock feedforward, the pipelined ADC uses two-phase non-overlapping clocks. There is a non overlapping phase in the clock of the sampling phase and the holding phase. Usually the width of the non overlapping phase is about 1/10 of the clock cycle. To improve the set-up time of an OPA in MDAC, it is necessary for the sub-ADC to compare and output the results in the non-overlapping clock phase. If the comparison time required by the comparator is greater than the non-overlapping clock phase, the time for OPA set-up will be reduced and the corresponding accuracy will decrease.

The resolution of sub-ADC is generally determined by the coarse quantization of each stage, that is, the ADC resolution in each stage should be equal to the effective conversion bit B_i. For a low-resolution flash ADC, its performance is mainly influenced by the comparator and the precision of the voltage reference. The effect of the two is equivalent to the offset voltage of comparator. In a flash ADC with resolution of B_i, the maximum allowable offset voltage of the comparator is:

$$V_{os,\max} = \pm \frac{V_{ref}}{2^{B_i}} \quad (6\text{-}22)$$

where V_{ref} is the amplitude of sub-ADC input. It can be seen that the requirement of comparator offset is relatively low in low-resolution flash ADC. Because the flash ADC used in pipelined ADC has low resolution, it creates conditions for the use of the dynamic comparator.

6.8.3 Comparator

In pipelined ADC, the sub-ADC is made up of a comparator. The function of comparator is to get the output by comparing the input signal and the reference signal. Fig. 6.25 is a functional schematic of an ideal comparator. When the input signal amplitude is less than the threshold, the comparator output is low voltage V_{OL}; and the comparator outputs a high voltage V_{OH} when the input signal amplitude is larger than the threshold. However, the comparator is not ideal, and there are non ideal factors such as limited gain, limited speed and offset. First, the limited gain determines the minimum signal amplitude that the comparator can distinguish. The larger the comparator gain, the smaller the signal amplitude that can be distinguished and the higher the comparator resolution. The limited speed determines whether it can be applied to high speed systems. The offset can be seen as a fixed error between the input signal and the threshold, and it also determines the scope of comparator application.

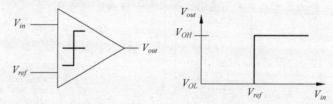

Fig. 6.25 The comparator function

The use of traditional OPAs provides better gain, but because of its limited speed, it is rarely used in high speed applications. At present, the structure which is commonly used is the pre-amplifier and latch structure as shown in Fig. 6.26.

The latch can be viewed as the loop with two inverters in series end to end, and its circuit model is shown in Fig. 6.27.

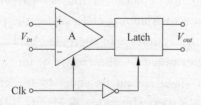

Fig. 6.26 Comparator structure

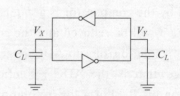

Fig. 6.27 Latch model

Fig. 6.28 is a small-signal model of the circuit. It can be obtained:

$$g_m V_X + \frac{V_Y}{r_0} + \frac{dV_Y}{dt}C_L = 0 \quad (6-23)$$

$$g_m V_Y + \frac{V_X}{r_0} + \frac{dV_X}{dt}C_L = 0 \quad (6-24)$$

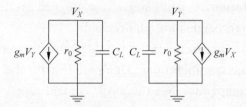

Fig. 6.28 Small-signal model of latch

where g_m, r_0 are the transconductor and output impendence of inverter respectively. Equation (6-23) and (6-24) are multiplied by r_0, then:

$$AV_X + V_Y = -\frac{dV_Y}{dt}\tau \quad (6-25)$$

$$AV_Y + V_X = -\frac{dV_X}{dt}\tau \quad (6-26)$$

$A = g_m r_0$ is the inverter gain, $\tau = r_0 C_L$ is the time constant. From (6-25) and (6-26), it can be get:

$$V_X - V_Y = -\frac{\tau}{A-1} \cdot \left(\frac{dV_X}{dt} - \frac{dV_Y}{dt}\right) \quad (6-27)$$

The initial value of the voltage difference between two X and Y is V_{XY0}, then it can be obtained:

$$V_X - V_Y = V_{XY0} \cdot e^{-\frac{t \cdot (A-1)}{\tau}} = V_{XY0} \cdot e^{-\frac{t}{\tau_{eff}}} \quad (6-28)$$

τ_{eff} is the equivalent time constant. It is assumed $A \gg 1$, then

$$\tau_{eff} = \frac{\tau}{A-1} \approx \frac{C_L}{g_m} \quad (6-29)$$

From equation (6-29), it is known that the smaller the equivalent time constant is, the faster the latch speed is. It means the output load C_L is as small as possible, and the inverter transconductance is as large as possible. If the equivalent time constant of the latch is larger than the design requirement, the latch will appear metastable, which requires a certain redundancy at the beginning of design.

In spite of the reasonable design, the latch can reach a high speed, but there are still some problems. First, because the effective input of the inverter is a certain range rather than the entire voltage range, and the input of the latch has a large offset, which affects the accuracy of comparator. Secondly, because the output of latch is a large signal, the output changes of latch are introduced into the input, resulting in a large kick-back noise. Kick-back noise not only affects the precision of the comparator, but also produces some interference to the voltage reference circuit. To solve the above two problems, a pre-amplifier is added to the comparator structure before the latch, which makes the offset and kick-back noise equivalent to the input effectively reduced, thus improving the comparator accuracy. The gain and bandwidth requirements of the pre-amplifier are determined by the design criteria of the comparator. In more strict design applications for bandwidth consideration, a multi-stage cascade pre-amplifier is usually used to achieve high gain and high bandwidth requirements.

In pipelined ADC, a comparator in its sub-ADC usually uses a dynamic comparator for power and precision considerations. Compared with the traditional comparator structure, the dynamic comparator has the clock control signal, so that the comparator does not consume current in half a clock cycle and has less power consumption.

6.8.4 MDAC

The main function of MDAC is to realize DAC conversion, sampling and holding, subtraction and gain amplification between stages. As the core module in pipelined ADC, the performance of MDAC is critical to the overall performance. Because the MDAC circuit will bring various errors, in order to reduce the effect of error to a reasonable range, MDAC structure with redundant bits is usually used. The transfer curve of most common used MDAC structure with 1.5bit is demonstrated in Fig.6.29.

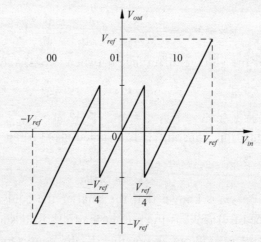

Fig.6.29 The transfer curve of MDAC structure with 1.5bit

The transfer curve of 1.5bit MDAC is divided into three intervals, which represent 00, 01, 10, respectively. This structure requires only two comparators to complete the conversion. There are two kinds of 1.5bit MDAC, one is the weight-capacitor structure, the other is the equal-capacitor structure. The structure of the weight-capacitor MDAC is shown in Fig.6.30. Fig.6.30(a) is the sampling phase. In sampling phase, the negative voltage reference is connected to C1, and the input signal is connected to the C2~C4. In amplification phase as shown in 6.30(b), C1 and C2 are connected to the output as the feedback capacitor, and the voltage reference with comparator output weight is added to the C3 and C4, respectively. The structure embodies the weight thought through the setting of C4 value.

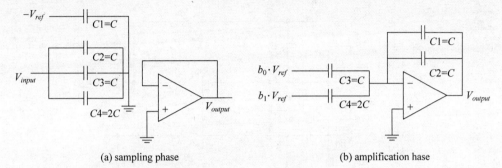

(a) sampling phase   (b) amplification hase

Fig. 6.30 1.5bit weight-capacitor MDAC

In sampling phase the charge in capacitor is:
$$Q_S = V_{in} \cdot 4C - V_{ref} \cdot C \tag{6-30}$$
In amplification phase the charge in capacitor is:
$$Q_h = 2C \cdot V_{out} + C \cdot b_0 \cdot V_{ref} + 2 \cdot C \cdot b_1 \cdot V_{ref} \tag{6-31}$$
b_1 and b_0 represent the high and low output codes of the comparator, respectively. Due to the principle of charge conservation, the charge of the sampling and holding phase is equal:
$$V_{in} \cdot 4C - V_{ref} \cdot C = 2C \cdot V_{out} + C \cdot b_0 \cdot V_{ref} + 2 \cdot C \cdot b_1 \cdot V_{ref} \tag{6-32}$$
then:
$$V_{out} = 2 \cdot V_{in} - \frac{1}{2} V_{ref} (2 \cdot b_1 + b_0) - \frac{1}{2} V_{ref} \tag{6-33}$$

In (6-33), the value of b_1 and b_0 is ± 1, and the piecewise function of the output can be obtained:
$$V_{out} = \begin{cases} 2 \cdot V_{in} - V_{ref} & V_{in} \geqslant V_{ref}/4 \\ 2 \cdot V_{in} & -V_{ref}/4 < V_{in} \leqslant V_{ref}/4 \\ 2 \cdot V_{in} + V_{ref} & V_{in} \leqslant -V_{ref}/4 \end{cases} \tag{6-34}$$

Thus the function of 1.5bit can be realized. The problem of this MDAC is the complexity of capacitor mismatch and coding. Due to the use of two capacitors 2C and C in the structure, the two kinds of capacitors bring more errors in layout matching. In addition, b_1 and b_0 in equation (6-33) can not be obtained directly from the flash ADC output, and additional coding is required. These two disadvantages make the structure less used in high-resolution pipelined ADC.

To reduce the mismatch of capacitor in layout, the more popular structure is the equal-capacitor MDAC. The structure is characterized by simple coding and equal capacitance value in the circuit, which makes the capacitor have better matching degree in the layout implementation. Its structure is shown in Fig. 6.31, where C_s is the sampling capacitor and C_f is the feedback capacitor with $C_s = C_f$. Φ_1 and Φ_2 are two-phase non-overlapping clocks, and Φ_{1a} changes to low voltage slightly before Φ_1 for bottom-plate sampling.

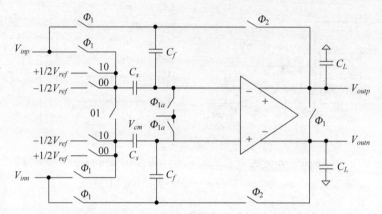

Fig. 6. 31　1. 5bit equal-capacitor MDAC

By deriving the output similar to 1. 5bit weight-capacitor MDAC, the output of 1. 5bit equal-capacitor MDAC can be obtained:

$$V_{out} = V_{in}\left(1 + \frac{C_s}{C_f}\right) - D \cdot \frac{C_s}{C_f} \cdot V_{ref} \tag{6-35}$$

Because the sampling capacitor C_s equals to feedback capacitor C_f, then:

$$V_{out} = 2V_{in} - D \cdot V_{ref} \tag{6-36}$$

The value of D is corresponding to the three intervals in Fig. 6.29, which are $-1, 0$, and $+1$, respectively.

6.9　Sigma-delta ADC

For the traditional Nyquist-sampling ADC, the mismatch degree of the components determines the resolution that the ADC can achieve. As the scaling of IC process, the matching error of components increases and the 2nd-order effect of MOS transistor is more and more significant. And the design of high-resolution Nyquist-sampling ADC faces more and more challenge. In the Nyquist-sampling ADC, the anti-aliasing transition band is very narrow, making the anti-aliasing filter complicated. To avoid these problems, the over sampling technique is used. First, under the over sampling condition, the sampling frequency of the signal is very high, which makes the requirement of the transition band of the anti-aliasing filter greatly reduced. The first or 2nd-order analog filter can meet the requirement. In addition, when designing high-resolution Nyquist-sampling ADC, because of the high accuracy match between components, we need to use complex laser trimming technique. By the use of over sampling technique, the requirements for components matching are also reduced.

Sigma-delta ADC is one of the most important innovations in the field of Analog IC design over the past few decades. For Sigma-delta modulator, it is a technology to improve the resolution of coarse quantizer by negative feedback, which was first proposed by Cutler in the 1960s. At the same time, a variant of the error feedback encoder is also proposed by the F. de Jager: the delta modulator. It is composed of a quantizer in the forward path and a loop filter in the feedback path. A few years later, Inose proposed to add a loop filter to the front of the delta modulator, and further improvements can move the loop filter into the feedback loop. If the loop filter is simplified as integrator, the system will include one integrator and one quantizer in the forward path, and the feedback loop includes a 1bit DAC. At this point, an delta modulator and an integrator are included in the system. In 1977, Ritchie made the first major improvement on the basic Sigma-delta modulator. He proposed that a series of integrators should be constructed in the forward path to form a higher-order loop filter, and feedback the output of DAC to the input of each integrator. In 1987, Lee proposed a stable high-order modulator design technique, that is, the Lee criterion. Based on this technique, the development of Sigma-delta modulator of the high-order loop filter with 4th-order above has been successfully developed. Then Hayashi proposed a cascade method to achieve a stable high-order Sigma-delta ADC, that is, Multi-stAge noise SHaping(MASH). The system first uses a single stage Sigma-delta modulator to process the input signal, and the quantization error is converted into a digital signal through a second stage Sigma-delta modulator. In digital output the quantization error of the first stage modulator is counteracted by a noise cancellation logic, and the quantization error of the second stage modulator is processed by noise shaping. This design method can be extended to high-order and multi-stagee converters, such as 3rd-stage(2-1 cascade), 4th-order(2-2 cascade, 2-1-1 cascade, etc.) MASH modulator design. In addition, the multi-bit internal quantizer technique can improve the performance of Sigma-delta modulator. This requires that a corresponding multi-bit DAC is included in the feedback loop. The DAC linearity limits the linearity of the whole Sigma-delta modulator. Carley uses a Dynamic Element Matching(DEM) method to reduce the nonlinear influence of multi-bit DAC.

Compared with the Nyquist-sampling ADC, over-sampling Sigma-delta ADC using over sampling and noise shaping technique will spread the thermal noise to the entire sampling spectrum, and push the quantization noise in the signal bandwidth to high frequency. Then the decimation filter is used to filter the quantization noise and achieve high resolution. The structure of Sigma-delta ADC is shown in Fig. 6.32.

Fig. 6.32 shows that the Sigma-delta ADC is composed of a Sigma-delta modulator and a digital decimation filter. The Sigma-delta modulator is mainly composed of a loop

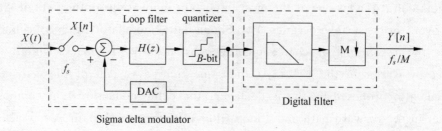

Fig. 6.32 The structure of Sigma-delta ADC

filter, a quantizer and a DAC. Sigma-delta modulator mainly completes signal over-sampling and quantization noise shaping. The decimation filter declines high-frequency quantization noise and down samples to Nyquist output.

6.9.1 Oversampling

In the Nyquist-sampling ADC, in order to prevent other signals from aliasing into the signal bandwidth, the sampling frequency should be greater than 2 times of the signal bandwidth. If the sampling frequency is much higher than Nyquist frequency, the noise power of signal bandwidth will decrease as the quantization error is uniformly distributed in the whole sampling frequency range.

As described above, within the range of its signal bandwidth the entire quantization noise power of the ADC can be expressed as

$$\sigma_{N,q}^2 = \frac{1}{f_s}\int_{-f_s/2}^{f_s/2} \frac{V_{LSB}^2}{12} df = \frac{V_{LSB}^2}{12} \tag{6-37}$$

We can get lower quantization noise by increasing the sampling frequency and making it far higher than the Nyquist frequency, as shown inequation (6-38).

$$\sigma_{O,q}^2 = \frac{1}{f_s}\int_{-f_b}^{f_b} \frac{V_{LSB}^2}{12} df = \frac{V_{LSB}^2}{12}\left(\frac{2 \cdot f_b}{f_s}\right) \tag{6-38}$$

From equation (6-38), it is known that when the sampling frequency f_s is far greater than the Nyquist frequency ($2f_b$), the quantization noise power in the signal bandwidth will fall according to the ratio of $f_s/2f_b$. The ratio of the sampling frequency to the Nyquist frequency is defined as the over sampling ratio (OSR), as shown in equation (6-39). And (6-38) can be expressed as (6-40). Using over sampling technique, the quantization noise power in the bandwidth can be reduced by OSR times.

$$OSR = f_s/2f_b \tag{6-39}$$

$$\sigma_{O,q}^2 = \frac{1}{f_s}\int_{-f_b}^{f_b} \frac{V_{LSB}^2}{12} df = \frac{V_{LSB}^2}{12}\left(\frac{2 \cdot f_b}{f_s}\right) = \frac{V_{LSB}^2}{12 \cdot OSR} = \frac{\sigma_{N,q}^2}{OSR} \tag{6-40}$$

Fig. 6.33 shows the power spectrum of quantization noise by over sampling technique.

According to equation (6-40), only considering the quantization noise, compared with the Nyquist-sampling ADC, the ideal SNR reached by over sampling ADC is

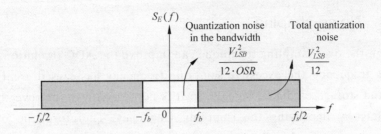

Fig. 6.33 The power spectrum of quantization noise by over sampling technique

$$SNR = P_{signal}/P_{noise} = (FS/2\sqrt{2})^2/(V_{LSB}^2/12) = 3 \cdot 2^{2N}/2 \quad (6\text{-}41)$$

and its logarithmic expression is

$$SNR_{dB} = 10\lg(P_{signal}/P_{noise}) = 6.02 + 1.76 + 10\lg(OSR) \quad (6\text{-}42)$$

Due to the use of over sampling technique, the effective resolution can be effectively improved. When the over sampling ratio is doubled, the ideal SNR increases about 3dB, and ENOB increases by about 0.5 bits.

The over sampling technique can also effectively reduce the transition band of the anti-aliasing filter. The main function of anti-aliasing filter is to filter the mirror signal that is aliasing in the signal band through the sampling process. Because the over sampling frequency is much higher than the Nyquist frequency, the sampled image signal is far away from the band signal, and the transition band of anti-aliasing filter is very wide.

Fig. 6.34(a) shows the requirement of an anti-aliasing filter for Nyquist-sampling ADC, and (b) is the requirement of an anti-aliasing filter for oversampling ADC. The transition band of anti-aliasing filter is usually required to be $f_{tb,n} = f_s - 2f_b$. Because the Nyquist frequency $2f_b$ of the Nyquist-sampling ADC is very close to the sampling frequency f_s, the transition band of the filter is very sharp. The sampling frequency of over-sampling ADC is far greater than the Nyquist frequency, so the transition band of the anti-aliasing filter is much wider and easier to be realized.

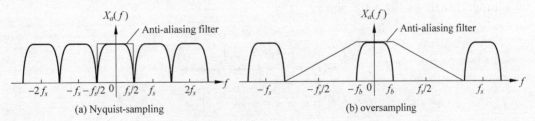

Fig. 6.34 The requirement of an anti-aliasing filter

The over sampling technique can not only effectively improve the ADC resolution, but also greatly reduce the transition band of the anti-aliasing filter and its design complexity. However, with a certain sampling frequency, increasing over sampling frequency is at the cost of reducing the effective bandwidth of signal. And because of the limitations of process and power consumption, the sampling frequency can not be infinite.

6.9.2 Noise shaping

Although the over sampling technique can improve the ADC resolution by increasing the sampling frequency, the excessive sampling frequency has caused great waste to the processing and storage of digital signals. So it is not realistic to improve the conversion accuracy solely by improving the sampling frequency. Therefore, the over sampling technique is generally combined with noise shaping to achieve the enhancement of ADC resolution. The basic idea of oversampling is to broaden the spectrum so as to "dilute" the noise in signal band. While the noise shaping technique is to push the noise of signal bandwidth to the high frequency band.

Noise shaping is a modulation scheme, which pushes the quantization noise outside the signal bandwidth in the form of high-pass filter. The higher order of the high-pass filter in the modulator and the greater the OSR, the smaller the noise power in the signal bandwidth. The linear model of the modulator is shown in Fig. 6.35, and its transfer function is shown in equation (6-43).

$$Y(z) = \frac{A(z)}{1+A(z)B(z)}X + \frac{1}{1+A(z)B(z)}\varepsilon_Q(z)$$
$$= STF(z) \cdot X(z) + NTF(z) \cdot \varepsilon_Q(z) \quad (6\text{-}43)$$

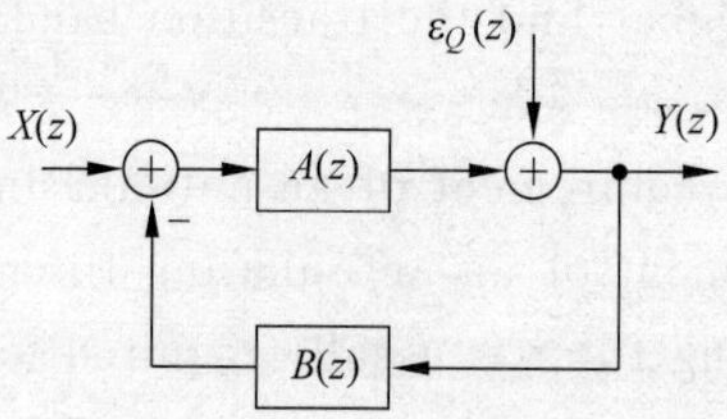

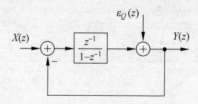

Fig. 6.35 The linear model of Sigma-delta modulator

In (6-43), STF is the Signal Transfer Function; NTF is the Noise Transfer Function. And $\varepsilon_Q(z)$ is quantization noise. Usually it is assumed $B(z) = 1$, so $A(z)$ is transferred into integral form as Fig. 6.36 shows.

Fig. 6.36 1st-order Sigma-delta modulator model

where $A(z) = \dfrac{z^{-1}}{1-z^{-1}}$ is integrator. Put it into (6-43), then:

$$Y(z) = z^{-1} \cdot X(z) + (1-z^{-1}) \cdot \varepsilon_Q(z)$$
$$= STF(z) \cdot X(z) + NTF(z) \cdot \varepsilon_Q(z) \quad (6\text{-}44)$$

In (6-44), $STF(z) = z^{-1}$, $NTF(z) = 1 - z^{-1}$. It can be seen that for the transfer function of 1st-order Sigma-delta modulator, the input signal $X(z)$ has only one clock cycle delay, and the quantization noise $\varepsilon_Q(z)$ is modulated by $NTF(z) = 1 - z^{-1}$.

It is assumed that the quantization noise is $E(n)$, and the difference between the quantization error of the adjacent two sampling is

$$E_1(z) = E(z) - E(z) \cdot z^{-1} = E(z)(1 - z^{-1}) \tag{6-45}$$

$E(z)$ is the quantization error of this sampling, and z^{-1} is the unit delay in frequency domain. The difference in quantization error between 2nd and 3rd-order Sigma-delta modulations is

$$E_2(z) = E(z) - 2E(z) \cdot z^{-1} + E(z) \cdot z^{-2} = E(z)(1 - z^{-1})^2 \tag{6-46}$$

$$E_3(z) = E(z) - 3E(z) \cdot z^{-1} + 3E(z) \cdot z^{-2} - E(z) \cdot z^{-3}$$
$$= E(z)(1 - z^{-1})^3 \tag{6-47}$$

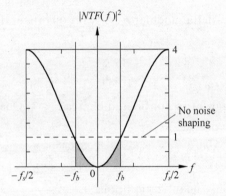

Fig. 6.37   1st-order noise shaping

In the same way, we can conclude the difference between the two adjacent quantization error of L-order Sigma-delta modulator:

$$E_L(z) = C_L^0 E(z) - C_L^1 E(z) \cdot z^{-1} + \cdots + (-1)^{(L-1)} C_L^{L-1} E(z) \cdot$$
$$z^{-(L-1)} + (-1)^L C_L^L E(z) \cdot z^{-L}$$
$$= E(z)(1 - z^{-1})^L \tag{6-48}$$

From the analysis of NTF in time domain, the difference between the quantization error of the adjacent two sampling is

$$NTF(\omega) = 1 - e^{-j\omega T} = 2j e^{-j\omega T/2} \frac{e^{j\omega T/2} - e^{-j\omega T/2}}{2j} = 2j e^{-j\omega T/2} \sin(\omega T/2) \tag{6-49}$$

It can be seen that the quantization noise is regarded as white noise, and the noise energy is shaped by the function $\sin^2(\omega T/2)$. The quantization noise is pushed to the high frequency, showing a more obvious attenuation at low frequency, as shown in Fig. 6.37. The total power of quantization noise within the bandwidth can be obtained:

$$V_n^2 = \varepsilon_Q^2 \int_0^{f_B} 4 \cdot \sin^2(\pi f T) df \approx \varepsilon_Q^2 \cdot \frac{4\pi^2}{3} f_B^3 T^2 \tag{6-50}$$

Its total energy can be expressed:

$$V_{n,Q}^2 = \varepsilon_Q^2 \cdot \frac{f_s}{2} \tag{6-51}$$

Put (6-51) into (6-50), The total power of quantized noise within the bandwidth can be obtained:

$$V_n^2 = V_{n,Q}^2 \cdot \frac{\pi^2}{3}\left(\frac{f_B}{f_s/2}\right)^3 = V_{n,Q}^2 \cdot \frac{\pi^2}{3} \cdot \frac{1}{OSR^3} \tag{6-52}$$

So, for the 1st-order Sigma-delta modulator of N bit quantizer, its ideal SNR satisfies the relationship:

$$SNR = 10 \cdot \lg \frac{P_s}{P_n} = (6.02N + 1.76) - 5.17 + 9.03 \cdot \log_2(OSR) \tag{6-53}$$

It can be concluded when OSR is increased by one time, SNR is improved about 9.03dB.

For the L-order Sigma-delta modulator of N bit quantizer, the NTF in frequency and time domain are:

$$NTF(z) = (1 - z^{-1})^L \tag{6-54}$$

$$|NTF(\omega)|^2 = |1 - e^{-j\omega T}|^{2L} = 2^{2L} \sin^{2L}(\omega T/2) \tag{6-55}$$

When OSR is very high, it is considered that $\omega T \ll 1$. Quantization noise power in signal bandwidth can be expressed as:

$$P_Q = \int_{-f_b}^{f_b} \frac{\Delta^2}{12 f_s} |NTF(f)|^2 df \approx \frac{\Delta^2}{12} \cdot \frac{\pi^{2L}}{(2L+1)OSR^{(2L+1)}} \tag{6-56}$$

The ideal SNR with only quantization noise is

$$SNR = \frac{P_S}{P_Q} = 3 \cdot 2^{2N-1} \cdot \frac{2L+1}{\pi^{2L}} OSR^{2L+1} \tag{6-57}$$

and the logarithmic form (6-58) is

$$SNR_{dB} = 10 \cdot \log_{10} \frac{P_S}{P_Q} = 6.02N + 1.76$$

$$+ 10 \cdot \log_{10}\left(\frac{2L+1}{\pi^{2L}} OSR^{2L+1}\right) \tag{6-58}$$

According to equation (6-58), L-order noise shaping combined with over sampling technique can increase ENOB of $L + 0.5$ bit when OSR is doubled, which is greatly improved compared to the 0.5 bits that only use over sampling technique.

We extend the model structure of Sigma-delta modulator to the general form, as shown in Fig.6.38. Fig.6.38(a) is the model of Sigma-delta modulator, and the Fig.6.38(b) is the linear model of the quantization noise for Sigma-delta modulator.

In 6.38(b) the transfer function of its negative feedback form can be derived as

$$[X(z) - Y(z)] \cdot H(z) + E(z) = Y(z) \tag{6-59}$$

$$Y(z) = \frac{H(z)}{1 + H(z)} X(z) + \frac{1}{1 + H(z)} E(z) \tag{6-60}$$

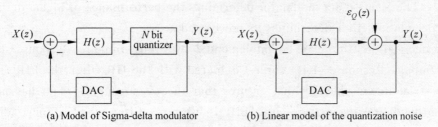

(a) Model of Sigma-delta modulator (b) Linear model of the quantization noise

Fig. 6.38 Model structure of Sigma-delta

The general form of STF and NTF can be obtained by comparing the equation (6-60) with (6-44):

$$STF(z) = \frac{H(z)}{1+H(z)} \tag{6-61}$$

$$NTF(z) = \frac{1}{1+H(z)} \tag{6-62}$$

If the loop filter $H(z)$ is a low-pass function, the gain in the low frequency of the signal bandwidth is large, while the gain in the high frequency is very small. Then the curve of $STF(f)$ and $NTF(f)$ with frequency is shown in Fig. 6.39. As can be seen from Fig. 6.39, $STF(f)$ is approximately a constant gain of 1 in the entire frequency band, that is, there is no effect on the input signal. $NTF(f)$ shows high-pass characteristics in the whole frequency band, and the noise in the signal bandwidth is suppressed, so the SNR will be greatly improved. The stronger the noise suppression ability, the more obvious the enhancement of the SNR.

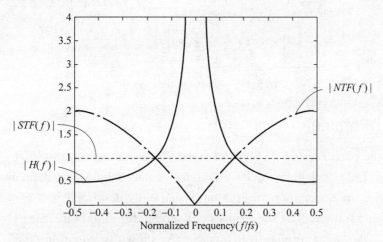

Fig. 6.39 The frequency curves of $H(f)$, $STF(f)$ and $NTF(f)$ for the 1st-order Sigma-delta modulator

6.9.3 Decimation filter

The decimation filter is an important part in Sigma-delta ADC. It mainly filters out the external noise of the signal baseband, and samples the output data down to Nyquist frequency. The Sigma-delta modulator determines the performance of Sigma-delta ADC, and the decimation filter determines its power and area.

The decimation filter can be implemented by Finite Impulse Response (FIR) or Infinite Impulse Response (IIR) filter. Compared with the IIR filter, the FIR filter can obtain a strict linear phase, which ensures that the over sampling data has no phase distortion by digital decimation. And FIR is all zeros filter which is unconditionally stable. While the IIR filter can only approximate to the linear phase. Therefore the decimation filter is usually realized by FIR filter.

Decimation filters are usually implemented in a multistage cascade. If a single stage is adopted, the order of the filter will be too high. So that the power consumption and the area are very large, and the hardware can not be realized. By using multistage FIR structure, the order of the filter can be reduced, thus the power and area of the filter can also be optimized. The comb filter is the simplest linear phase FIR filter because it does not need the multiplier. The comb filter is usually used as the first stage of the decimation filter, and large decimation factor is achieved. Since the comb filter has a certain amplitude attenuation in the passband, it is necessary to compensate the filter for a certain amount with compensation filter. The last stage is a half-band filter, which is used to get a very sharp transition zone. A schematic of a cascade decimation filter is shown in Fig. 6.40.

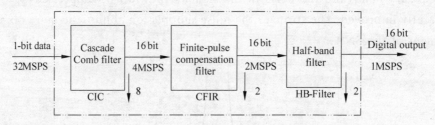

Fig. 6.40 Cascade decimation filter

6.9.4 Parameter

Because the performance of Sigma-delta ADC is mainly determined by Sigma-delta modulator, the following Sigma-delta ADC are replaced by Sigma-delta modulator for simplicity, while the decimation filter is not specifically described.

The design parameters of the Sigma-delta modulator mainly include: Over Sampling Ratio(OSR), order(L) and quantizer bits(N). As shown in equation (6-56), the higher order Sigma-Delta modulator can better shape the quantization noise to achieve higher conversion resolution. Different order modulators have different NTF. When the quantizer bit is 1, the relationship between the different order modulator and the NTF is shown in Fig. 6.41.

In Fig. 6.41, Compared with the 1st-order Sigma-Delta modulator, higher order modulator NTF further reduces the quantization noise in low frequency band, and amplifies

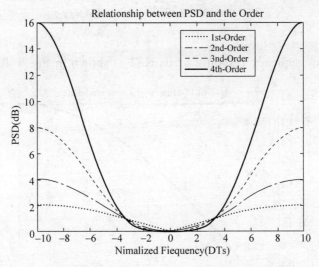

Fig. 6.41 The relationship between the different order modulator and the NTF

quantization noise in high frequency. That is, the quantization noise is further pushed to the higher frequency. The higher the order is, the stronger the noise suppression ability is and the more obvious the effect is.

6.9.5 The basic structure of Sigma-delta modulator

1. Low-order Sigma-delta modulator

1) 1st-order Sigma-delta modulator

Fig. 6.42 shows the structure of the 1st-order Sigma-delta modulator, in which the loop filter is implemented by a 1st-order integrator.

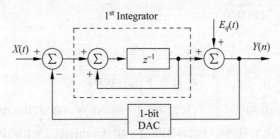

Fig. 6.42 1st-order Sigma-delta modulator

The STF and NTF of the 1st-order Sigma-delta modulator are:

$$STF(z) = \frac{H(z)}{1+H(z)} = z^{-1} \qquad (6\text{-}63)$$

$$NTF(z) = \frac{1}{1+H(z)} = 1 - z^{-1} \qquad (6\text{-}64)$$

From (6-63) and (6-64) we know that the signal only has a cycle delay, and the noise is shaped by the 1st-order. When $z = e^{j2\pi f/f_s}$, in time domain the STF and NTF are

$$NTF(f) = |1 - e^{-j2\pi f/f_s}| = 2\sin(\pi f/f_s) \tag{6-65}$$

$$STF(f) = |e^{-j2\pi f/f_s}| = 1 \tag{6-66}$$

The amplitude-frequency response of its NTF is shown in Fig.6.43.

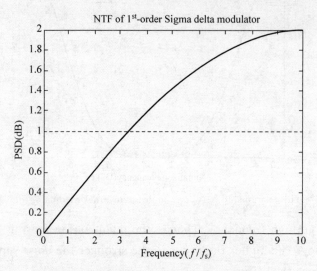

Fig. 6.43　The amplitude-frequency response of 1st-order modulator NTF

The quantization noise power in the bandwidth of the 1st-order Sigma-delta modulator and the ideal SNR can be derived as:

$$P_n = \int_{-f_b}^{f_b} \frac{\Delta^2}{12} \times \frac{1}{f_s} |NTF(z)|^2 df = \int_{-f_b}^{f_b} \frac{\Delta^2}{12} \times \frac{1}{f_s} [2\sin(\pi f/f_s)]^2 df$$

$$= \frac{\Delta^2 \pi^2}{36} \frac{1}{OSR^3} \tag{6-67}$$

$$SNR|_{max} = 10\lg \frac{P_s}{P_n} = 10\lg\left(\frac{3}{2} 2^{2N}\right) + 10\lg\left[\frac{3}{\pi^2}(OSR)^3\right]$$

$$= 6.02N + 1.76 - 5.17 + 30\lg(OSR) \tag{6-68}$$

As shown in equation (6-67), for the 1st-order Sigma-delta modulator, when OSR increases by 1 times, the SNR increases about 9dB. Compared with the non noise shaping, the SNR has been improved effectively.

As shown in Figure 6.44, the 1st-order Sigma-delta modulator is composed of a sampling circuit, an integrator, a quantizer, and a 1 bit feedback DAC.

2) 2nd-order Sigma-delta modulator

Increasing the NTF order of Sigma-delta modulator can effectively reduce the quantization noise in the signal bandwidth. Figure 6.45 shows the structure of the 2nd-order Sigma-delta modulator, in which the loop filter is implemented with a 2nd-order integrator.

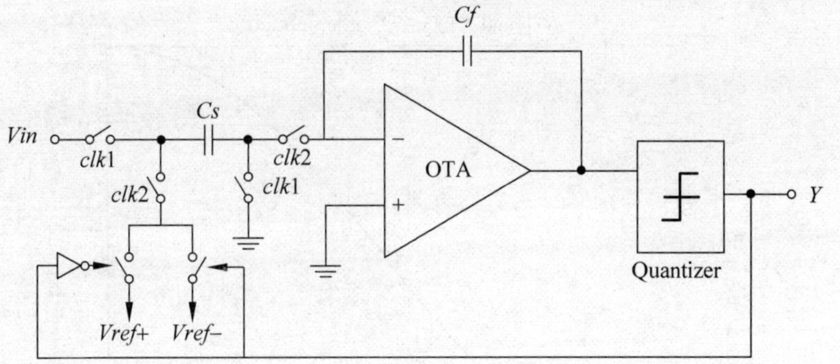

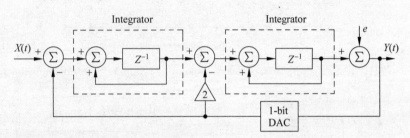

Fig. 6.44　The implementation of 1st-order Sigma-delta modulator

Fig. 6.45　2nd-order Sigma-delta modulator

Its STF and NTF are:

$$STF(z) = \frac{H^2(z)}{1 + 2H(z) + H^2(z)} = z^{-2} \qquad (6\text{-}69)$$

$$NTF(z) = \frac{1}{1 + 2H(z) + H^2(z)} = (1 - z^{-1})^2 \qquad (6\text{-}70)$$

For (6-69) and (6-70) it is known that the signal has two cycle delay, and the noise is shaped by the 2nd-order. Set $z = e^{j2\pi f/f_s}$, the STF and NTF in time domain are:

$$STF(f) = |\,e^{-j2\pi f/f_s}\,|^2 = 1 \qquad (6\text{-}71)$$

$$NTF(f) = |\,1 - e^{-j2\pi f/f_s}\,|^2 = [2\sin(\pi f/f_s)]^2 \qquad (6\text{-}72)$$

The amplitude-frequency response of its NTF is shown in Fig. 6.46.

Its ideal SNR is:

$$SNR\,|_{max} = 10\lg \frac{P_s}{P_n} = 6.02N + 1.76 - 12.9 + 50\lg(OSR) \qquad (6\text{-}73)$$

For the 2nd-order Sigma-delta modulator, when OSR increases by 1 times, and the SNR increases about 15dB. Compared with the 1st-order noise shaping, its SNR has been improved further. The 2nd-order Sigma-delta modulator consists of a sampling circuit, a two-stage integrator, a quantizer, and a 1 bit DAC as shown in Fig. 6.47.

When the NTF is further increased, the quantization noise power in the signal bandwidth can be reduced more, and the performance can also be further improved.

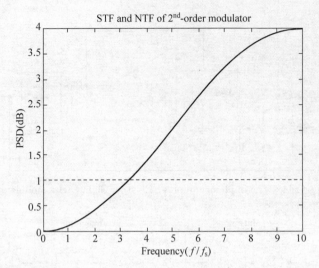

Fig. 6.46　The NTF amplitude-frequency response of 2nd-order Sigma-delta modulator

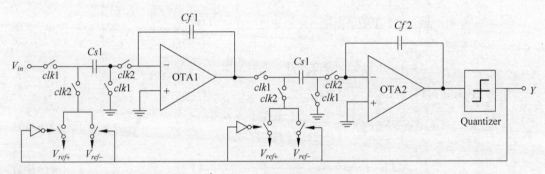

Fig. 6.47　2nd-order Sigma-delta modulator structure

Usually, its NTF can be expressed as equation (6-74), and the quantization noise power and ideal SNR in signal bandwidth are rewritten as (6-75) and (6-76).

$$NTF(z) = (1 - z^{-1})^L \tag{6-74}$$

$$P_Q = \int_{-f_b}^{f_b} \frac{\Delta^2}{12 f_s} \mid NTF(f) \mid^2 df \approx \frac{\Delta^2}{12} \cdot \frac{\pi^{2L}}{(2L+1) OSR^{(2L+1)}} \tag{6-75}$$

$$SNR_{dB} = 10 \cdot \lg \frac{P_S}{P_Q} = 6.02N + 1.76 + 10 \cdot \lg \left(\frac{2L+1}{\pi^{2L}} OSR^{2L+1} \right) \tag{6-76}$$

2. Single-loop high-order Sigma-delta modulator

All integrators of a single-loop high-order Sigma-delta modulator are in the same feedback loop, as shown in Fig.6.48. Its advantage is that it can achieve high SNR. The circuit structure is simple, and it is not sensitive to the non ideal characteristics of integrator and quantizer. But its disadvantages are also very obvious. Because the integrator is in the same loop, when the order is higher, the gain of the high frequency of the cascade integrator transfer function is obviously increased, resulting in the instability of the whole system.

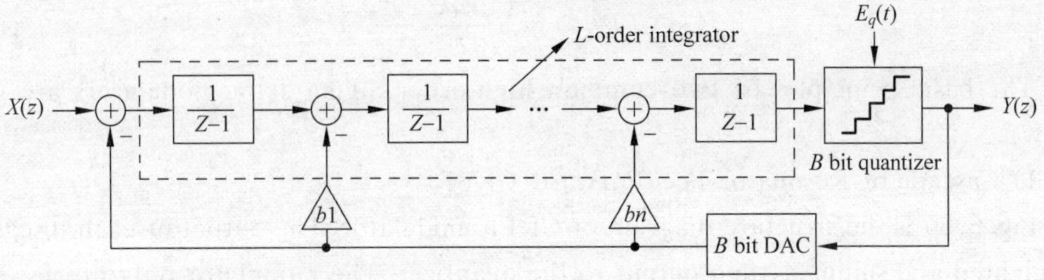

Chapter 6 Analog-to-Digital Converter

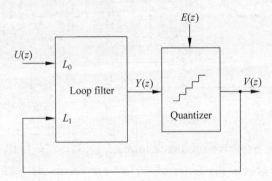

Fig. 6.48 Single-loop high-order Sigma-delta modulator

As described in the previous section, the transfer function of Sigma-delta modulator can be divided into two functions: the signal transfer function STF and the noise transfer function NTF. Usually, the Sigma-delta modulator is consists of two parts: loop filter (linear part) and quantizer(non-linear part). The single-ended output can be represented as a linear combination of two inputs, as shown in Fig. 6.49. And its expression are (6-77) and (6-78).

$$Y(z) = L_0(z)U(z) + L_1(z)V(z) \qquad (6\text{-}77)$$

$$V(z) = Y(z) + E(z) \qquad (6\text{-}78)$$

Fig. 6.49 General structure of Sigma-delta modulator

From (6-77) and (6-78), the expression of $V(z)$ can be obtained:

$$V(z) = STF(z)U(z) + NTF(z)E(z) \qquad (6\text{-}79)$$

and the STF and NTF are

$$NTF(z) = \frac{1}{1 - L_1(z)}, \quad STF(z) = \frac{L_0(z)}{1 - L_1(z)} \qquad (6\text{-}80)$$

For different modulators, $L_0(z)$ and $L_1(z)$ of loop filters can represent system functions of different parameters. As the order of Sigma-delta modulator is increasing, the expressions of $L_0(z)$ and $L_1(z)$ will become more and more complex. $L_1(z)$ has a high gain within the signal bandwidth and attenuates the quantization noise sufficiently. Because NTF determines the noise suppression ability and system stability, in general, the design of Sigma-delta modulator starts from the design of NTF.

In order to get a high-order stable Sigma-delta modulator, it is necessary to select the appropriate pole, making the NTF is:

$$NTF(z) = \frac{(1-z^{-1})^L}{D(z)} \tag{6-81}$$

The basic principles of two common high-order Sigma-delta modulators are given below:

1) Cascade of Resonator FeedForward(CRFF)

Fig. 6.50 is the structure diagram of CRFF modulator. The output of each integrator is weighted and summed, then output to the quantizer. The modulator only processes the signal of former integrator. Or when there is a direct path from the input to the quantizer, all integrators do not process the input signal, but only deal with the quantization noise, which directly reduces the output swing requirement of the integrator. Its transfer function and STF are:

$$L_0(z) = -L_1(z) = \frac{a_1}{z-1} + \frac{a_2}{(z-1)^2} + \frac{a_3}{(z-1)^3} + \cdots + \frac{a_n}{(z-1)^n} \tag{6-82}$$

$$STF(z) = 1 - NTF(z) \tag{6-83}$$

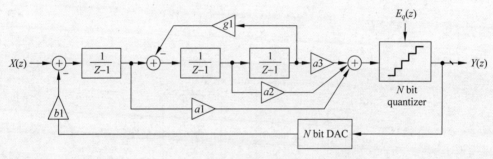

Fig. 6.50 CRFF modulator

In Fig. 6.50, if the negative feedback loop of g_1 is not included, the pole of $L_1(z)$ is limited to DC point. Because the poles of $L_1(z)$ are zeroes of NTF, all zeros of NTF are at DC point. By adding g_1, the transfer function of modulator forms a resonator, which moves the poles out of DC point. And NTF's zeros are moved from the DC range to the signal bandwidth, which can better suppress the quantization noise and achieve better performance.

2) Cascade of Resonator FeedBack(CRFB)

Fig. 6.51 is a structure diagram of the CRFB modulator. The input of each integrator has the negative feedback of the output for differential operation. And its transfer function satisfies:

$$L_0(z) = \frac{b_1}{(z-1)^n} \tag{6-84}$$

$$-L_1(z) = \frac{a_1}{z-1} + \frac{a_2}{(z-1)^2} + \frac{a_3}{(z-1)^3} + \cdots + \frac{a_n}{(z-1)^n} \tag{6-85}$$

In Fig. 6.51, if the negative feedback loop of g_1 is not included, all the zero of the NTF are at the DC point. The NTF determines $L_1(z)$ and STF. When the NTF is

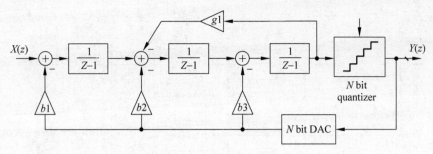

Fig. 6.51 CRFF modulator

expressed as equation (6-86), the STF is obtained as equation (6-87).

$$NTF(z) = \frac{(z-1)^n}{D(z)} \tag{6-86}$$

$$STF(z) = NTF(z)L_0(z) = \frac{b_1}{D(z)} \tag{6-87}$$

After adding g_1, the transfer function of the modulator forms a resonator, and its expression is

$$R(z) = \frac{z}{z^2 - (2-g)z + 1} \tag{6-88}$$

If the appropriate feedback coefficient and feedforward coefficient are selected for the CRFF and CRFB structures, the equivalent NTF can be obtained. However, their STF are not the same. In CIFF structure, because the signal is directly feedforward to the output after the first integrator, the STF has 1st-order filter characteristic. While in the CIFB structure, the signal passes through all integrators to the output, therefore its STF has the L-order filter feature.

3. Cascaded high-order Sigma-delta modulator

Because the structure of single-loop high-order Sigma-delta modulator is more complex, it can not be analyzed by linear system. Meanwhile the stability of the system needs special consideration. So the new high-order Sigma-delta modulator structure is generated.

Then a cascade of low-order modulators structure is adopted, and each stage contains only 1st or 2nd-order integrators. The structure uses the first-stage quantization noise as the input of the post-stage modulator. At the same time a output noise cancellation logic is added to cancel the quantization noise of the former stage, leaving only the input signal and the last stage quantization noise after noise shaping. This structure is called a Multi-stAge noise SHaping(MASH) modulator. Fig. 6.52 is a structure diagram of a 2-1 cascaded MASH modulator, which can actually reach the 3rd-order noise shaping capability. The output of the first and second modulators is shown as (6-89) and (6-90), respectively.

$$Y_1(z) = z^{-2}X(z) + (1-z^{-1})^2 Q_1(z) \tag{6-89}$$

$$Y_2(z) = z^{-1}g_1 Q_1(z) + (1-z^{-1})Q_2(z) \tag{6-90}$$

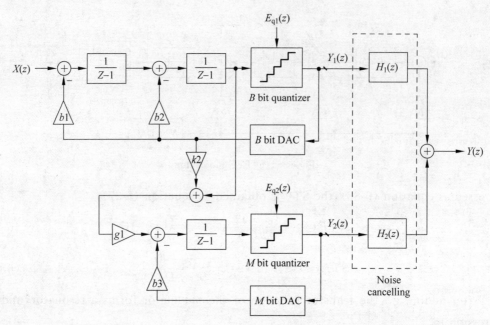

Fig. 6.52 MASH 2-1 modulator

where $X(z)$ is the input signal. $Q_1(z)$ and $Q_2(z)$ are the quantization noise of the first and the second stage modulator, respectively. In addition, k_2 is the output weight coefficient to the next stage. g_1 is the scaling factor of quantization noise. The output of the two modulator are $Y_1(z)$ and $Y_2(z)$. The quantization noise of the first modulator is canceled by noise cancellation logic $H_1(z)$ and $H_2(z)$, so that the final output $Y_1(z)$ contains only the quantization noise of the second modulator, and achieves 3rd-order noise shaping function. The final transfer function is

$$Y(z) = Y_1(z)H_1(z) + Y_2(z)H_2(z) \qquad (6-91)$$

And the noise cancellation logic are

$$H_1(z) = z^{-1} \qquad (6-92)$$

$$H_2(z) = \frac{1}{g_1}(1-z^{-1})^2 \qquad (6-93)$$

Combined with (6-89) to (6-93), the transfer function of MASH 2-1 is

$$Y(z) = z^{-3}X(z) + \frac{1}{g_1}(1-z^{-1})^3 Q_2(z) \qquad (6-94)$$

As shown in equation (6-94), the quantization noise of the first stage is completely eliminated, and the quantization noise of the second stage is processed by 3rd-order noise shaping. Each stage of the modulator is low-order structure without the stability problem.

Since equation (6-89) to (6-93) are performed in analog domain, and (6-94) is completed in digital domain. Two domains are completely different in operation. The implementation of analog domain is the ratio of capacitance, which depends on the matching accuracy of integrated circuit technology, and there is a certain error; While the realization of the digital domain is the shift and addition operation, and there is no error ideally.

There are more or less errors between analog and digital domain. So the mismatch caused by this error will make the quantization noise of the former stage not be completely canceled, resulting in the quantization noise leakage to the output, and ultimately the quality of output signal will decrease.

In order to reduce the mismatch, the MASH modulator is usually implemented in the CMOS process with higher capacitor matching accuracy and switched-capacitor structure. Besides this mismatch can also be reduced by increasing the capacitor area and the gain bandwidth product of the integrator in circuit design.

4. Multibit Sigma-delta modulator

From the ideal SNR equation of Sigma-delta modulator, the SNR is related to the over sampling ratio OSR, the modulator order L and the quantizer bit N. To improve SNR, it is necessary to increase the OSR, L, or N.

To improve OSR means improving the sampling frequency. When the signal bandwidth reaches MHz orders of magnitude, only increasing the sampling frequency, on the one hand, makes the power consumption of the circuit increase dramatically; and on the other hand, it will not be realized due to the limited technological conditions. Since the Sigma-Delta modulator is a nonlinear negative feedback closed-loop system, when the modulator order L is greater than two, it will cause the system instability, making the quantizer overload and the performance go down rapidly. A more appropriate way is to improve the performance by increasing the quantizer bit N. Moreover, increasing N will enhance the stability of the higher-order modulator, and increase the stable input swing of the quantizer. Fig. 6.53 is the relationship between the quantizer bit N of 3rd-order modulator and the peak SNR.

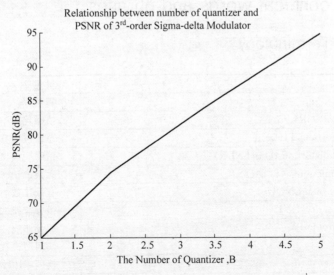

Fig. 6.53 The relationship between the quantizer bit N of 3rd-order modulator and the peak SNR

However, if the modulator uses a multibit quantizer, the multibit DAC will be used in the feedback loop, and the accuracy of the DAC has a great impact on the modulator. Taking 1st-order multibit modulator as an example, $X(z)$ is the input signal, and $E_Q(z)$ is the quantization noise. $E_D(z)$ is the noise introduced by the nonlinear error of feedback DAC, and its transfer function is.

$$Y(z) = z^{-1}X(z) + (1-z^{-1})E_Q(z) - z^{-1}E_D(z) \qquad (6-95)$$

It can be seen that the nonlinear error generated by multibit DAC is not modulated by the feedback loop as quantization noise, so the resolution of the whole modulator is limited by the multibit feedback DAC.

In order to solve the nonlinear problem of multibit DAC, many DAC linearization techniques and methods have been proposed, and the more practical is the Data Weighted Averaging (DWA) algorithm. In this algorithm, the number of times each data is selected is basically equal, then the difference between each data is averaged. The basic principle is to use a pointer to locate. After each selection, the pointer is positioned at the end of the unit sequence. Therefore, in the next selection we choose the first element that is not used. Fig. 6.54 is the selection order of a 3 bit DAC. The transverse number represents the number of capacitor (a total of 7). The vertical number represents the number of capacitor selected each time (6 times). And the black shadow area of each row represents the selected capacitors.

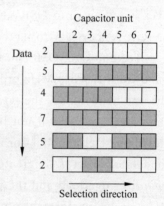

Fig. 6.54　DWA algorithm

6.10　Technical words and phrases

6.10.1　Terminology

quantizer	量化器
Full Scale (FS)	满量程
signal-to-noise ratio (SNR)	信噪比
total harmonic distortion (THD)	总谐波失真
signal-to-noise and distortion ratio (SNDR)	信噪失真比
spurious-free dynamic range (SFDR)	无杂散动态范围
effective number of bits (ENOB)	有效位数
Differential Nonlinearity (DNL)	微分非线性
Integral Nonlinearity (INL)	积分非线性
flash ADC	快闪型模数转换器
Flash-interpolation structure	快闪-插值结构
Folding-interpolation structure	折叠-插值结构

续表

variable gain amplifier(VGA)	可变增益放大器
sample and hold amplifier(SHA)	采样保持放大器
Successive Approximation Analog-to-digital Converter(SAR ADC)	逐次逼近模/数转换器
digital-to-analog converter(DAC)	数模转换器
Pipelined ADC	流水线模数转换器
Multiplying DAC(MDAC)	乘法数模转换器
Multi-stAge noise SHaping(MASH)	多级噪声整形
over sampling ratio(OSR)	过采样率
anti-aliasing filter	抗混叠滤波器
decimation filter	降采样滤波器
finite impulse response(FIR)	有限冲击响应滤波器
infinite impulse response filter(IIR)	无限冲击响应滤波器
Cascade of Resonator FeedForward(CRFF)	级联前馈谐振器
Cascade of Resonator FeedBack(CRFB)	级联反馈谐振器
Data Weighted Averaging(DWA) algorithm	数据加权平均算法

6.10.2　Note to the text

(1) ADC is a circuit unit that converts analog signals into digital signals. Its main function is to convert a continuous analog signal in time and amplitude into a digital signal that is discrete in time and amplitude.

模/数转换器是一种将模拟信号转换为数字信号的电路单元。它的主要功能是将在时间和幅度上都连续的模拟信号转换为在时间和幅度上都是离散的数字信号。

(2) The folding and interpolation structure combines two technical advantages, which reduce the input capacitor, chip area and power consumption, and the dynamic performance is also improved.

折叠-插值结构结合了两种技术优势，降低了输入电容、芯片面积和功耗，而且提升了动态性能。

(3) The principle of folding ADC is to use preprocessing technique to reduce hardware consumption and to maintain one-step conversion characteristic of flash ADC.

折叠式 ADC 的原理是用预处理技术减小硬件消耗，同时保持快闪式 ADC 的一步转换特性。

(4) Interpolation technique uses the linear characteristics of flash ADC near the threshold voltage, through resistor series, current mirrors, capacitors to generate more voltage reference.

内插技术就是利用快闪式 ADC 在阈值电压附近的线性特性，通过电阻串、电流镜、电容等内插快闪式 ADC 所需的更多的参考电压。

(5) But the interpolating structure alone can not reduce the number of latches. Therefore, interpolating ADC is usually combined with folding technique, so as to reduce the number of pre-amplifiers and latches at the same time, and further reduce the area and power consumption.

但是单独使用插值结构并不能减少锁存器的数量,因此通常结合折叠技术构成折叠-插值 ADC,从而同时减小前置运放和锁存器数目,进一步减小面积与功耗。

(6) In addition to the influence of the on-resistance on the sampling accuracy, the structure also has a more serious clock feedthrough and charge injection effect.

除了开关的导通阻抗对采样精度的影响外,该结构还存在比较严重的时钟馈通和电荷注入效应。

(7) First, because the effective input of the inverter is a certain range rather than the entire voltage range, and the input of the latch has a large offset, which affects the accuracy of comparator.

首先,由于反相器的有效输入是一个范围而非整个电压区间,因此锁存器的输入端存在着较大的失调,这影响着比较器的精度。

(8) Compared with the Nyquist-sampling ADC, over-sampling Sigma-delta ADC using over sampling and noise shaping technique will spread the thermal noise to the entire sampling spectrum, and push the quantization noise in the signal bandwidth to high frequency.

相对于 Nyquist 采样频率 ADC,过采样 Sigma-deltaADC 采用过采样和噪声整形技术将热噪声平铺至整个采样频谱内,并将信号带宽内的量化噪声推向高频。

(9) Because the over sampling frequency is much higher than the Nyquist frequency, the sampled image signal is far away from the band signal, and the transition band of anti-aliasing filter is very wide.

由于过采样的采样频率远高于 Nyquist 频率,所以其采样后的镜像信号距离带内信号很远,所以对过采样 ADC 的抗混叠滤波器的过渡带就比较宽。

(10) The higher order of the high-pass filter in the modulator and the greater the OSR, the smaller the noise power in the signal bandwidth.

调制器中的高通滤波器的阶数越高、过采样率越大,信号带宽内的噪声功率就越小。

(11) Because the integrator is in the same loop, when the order is higher, the gain of the high frequency of the cascade integrator transfer function is obviously increased, resulting in the instability of the whole system.

由于积分器在同一个环路内,当阶数较高时,级联积分器传输函数的高频段增益明显增大,导致整个系统不稳定。

Part III

Part II

Chapter 7

Low Power Design for Digital CMOS Circuits

In recent years, with the development of wearable devices, low power design has become an important challenge for CMOS integrated circuits. In order to extend the operation time of device, the power consumption of the core chip must be within a reasonable range. Before starting introduction, it is necessary to differentiate power and energy first, especially for battery operated system. Power is the instantaneous power dissipation in the system, and energy is the integral of power over time. The power used by a given system varies over time depending on what it is doing, while it is energy that determines battery life.

At present, CMOS is the mainstream technology for digital VLSI. In this section we firstly introduce the main source of power consumption in CMOS circuits, and then give a briefly summary on low power design Methodologies.

7.1 Sources of Power Consumption

There are a number of sources of power consumption in digital CMOS circuits, which can be subdivided into dynamic and static power consumption. Dynamic power is the power consumed when the device is active, which means when signals are changing values. Static power is the power consumed when the device is powered up but no signals are changing value.

1. Dynamic Power

Dynamic power consumption mainly consists of switching power and short circuit power, of which the former dominates. Switching power, known as capacitive power, is consumed during charging and discharging the parasitic load capacitor.

Fig. 7.1 illustrates switching power by the transition of a basic CMOS inverter.

The energy required per transition in a CMOS inverter is a fixed value, given by:

$$E = C_L \cdot V_{DD}^2 \tag{7-1}$$

Where C_L is the load capacitance and V_{DD} is the supply voltage. If the frequency of the system clock is f, and there are N nodes in the system, then the switching power of the system can be modeled by the following equation:

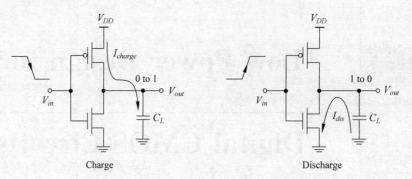

Fig. 7.1 Switching power of a CMOS inverter

$$P_{switch} = V_{DD}^2 \cdot f \cdot \sum_{i=1}^{N} \alpha_i \cdot C_i \tag{7-2}$$

Where C_i and α_i is the load capacitor and probability of output transition for node i. If we define:

$$C_{eff} = \sum_{i=1}^{N} \alpha_i \cdot C_i \tag{7-3}$$

We can also describe the switching power as below:

$$P_{switch} = C_{eff} \cdot V_{DD}^2 \cdot f \tag{7-4}$$

The power dissipation due to short-circuit current is caused by the inherent non-ideal characteristic of the input signals in CMOS circuits. Because of the finite slope of the input signal during switching, the PMOS and NMOS devices are simultaneously turned ON for a very short period of time during a logic transition, allowing a short-circuit current to run from V_{DD} to ground. Fig. 7.2 shows the short-circuit current in a CMOS inverter.

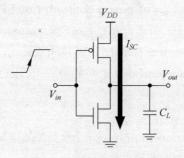

Fig. 7.2 Short circuit current of a CMOS inverter

Normally, short-circuit power is only a small part in dynamic power. At the overall circuit level, if rise/fall times of all signals are kept constant within a range, overall short-circuit current will be kept within bounds (maximum 10-15% of the total dynamic power dissipation). In the extreme case, if $V_{DD} < VTHn + |VTHp|$, the shortcircuit dissipation will be completely eliminated, because the devices are never ON simultaneously.

For the overall dynamic power is dominated by the switching power, we often simply use the switching power formula to compute the total dynamic power:

$$P_{dyn} \approx P_{switch} = C_{eff} \cdot V_{DD}^2 \cdot f \qquad (7\text{-}5)$$

Formula (7-5) is very important for power computation for CMOS digital integrated circuits. From the (7-5), we can draw conclusion that switching power is not a function of transistor size, resistance, charging/discharging current, but rather a function of supply voltage, clock frequency, switching activity and load capacitor. To reduce the dynamic power dissipation, there are many techniques can be used, which focus on the voltage, frequency, capacitor components of the equation, as well as reducing the data-dependent switching activity.

2. Static Power

In digital CMOS circuits, static power, which is a constant factor and has nothing to do with the switching activity, is mainly caused by leakage current. There are four main sources of leakage currents in a CMOS device: Sub-threshold Leakage (I_{SUBTH}), Gate Leakage (I_{GATE}), Diode (drain-substrate) Reverse Bias Junction Leakage (I_{REV}), Gate Induced Drain Leakage (I_{GIDL}). Fig. 7.3 illustrates the main leakage currents in a MOS device.

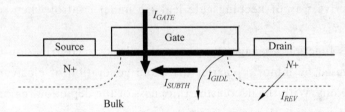

Fig. 7.3 Main leakage current in MOS device

Among all the above listed leakage current components, sub-threshold leakage is the dominant component of leakage current in previous technology nodes. Subthreshold leakage current flows from the drain to the source of a transistor when it is off (in fact not turned completely off). This happens in weak inversion mode when the applied voltage V_{GS} is less than the threshold voltage V_{TH} of the transistor. The value of sub-threshold leakage (I_{SUBTH}) current can be given by the following equation:

$$I_{SUBTH} = \mu C_{ox} V_{T0}^2 \frac{W}{L} \cdot e^{\frac{V_{GS}-V_{TH}}{nV_{T0}}} \qquad (7\text{-}6)$$

Where W and L are the dimensions of the MOS transistor, the n is a function of the device fabrication process and ranges from 1.0 to 2.5, and V_{T0} is the thermal voltage kT/q.

Sub-threshold leakage makes low power design rather complicated. For reducing dynamic power, people choose to decrease supply voltage V_{DD} continuously. Meanwhile, in order to maintain the good performance, we need to lower threshold voltage V_{TH} as we lower V_{DD}. But unfortunately, lowering V_{TH} will result in an exponential increase in the sub-threshold leakage current according to equation (7-6). Even worse, sub-threshold leakage current increases exponentially with temperature, which makes the leakage power easy to exceed the design constraint at the worst case while it is acceptable at room temperature.

There are several methods to deal with sub-threshold leakage current, one of them is using multi-threshold logic. For critical path in the circuit, low threshold cells are used to achieve the timing constraint, while for other paths, high threshold cells with lower sub-threshold leakage can be chosen. Another important technique is power-gating, which shut down the supply voltage parts of the chip when they are not working. Some other methods often used for lowering sub-threshold leakage are variable threshold CMOS, utilizing the stack effect, long channel devices, and so on.

Gate leakage, which occurs as a result of tunneling current through the gate oxide, is becoming significant in the sub-100 nm era. To maintain the current drive of the MOS transistor while scaling its horizontal dimensions, the gate oxide(SiO2) thickness is scaled as well. When the oxide thickness becomes of the order of just a few molecules in the sub-100 nm technology nodes, gate tunneling current can increase to the amount that may be comparable with sub-threshold leakage. Now, using high-k dielectric materials appears to be the only effective way of keeping gate leakage under control when scaling the gate thickness.

Reverse Bias Junction Leakage(I_{REV}), another leakage contribution that should not be ignored, is caused by minority carrier drift and generation of electron/hole pairs in the depletion regions. With the decreasing thickness of the depletion regions owing to the high doping levels, tunneling through narrow depletion region becomes an issue in sub-50 nm technology nodes.

Gate Induced Drain Leakage(I_{GIDL}) is the current flowing from the drain to the substrate induced by the high electric field effect in the MOSFET drain caused by a high V_{DG}. Generally speaking, I_{GIDL} plays a minor role and can be ignorable in most of today's designs.

Above 90 nm technology node, static power in digital CMOS circuits is usually very small compared to dynamic power. But in the sub-90 nm technology nodes, static power has become a big problem and should be paid much attention to, especially for the battery operated systems widely used today.

7.2 Low Power Design Methodologies

Low-power design can be applied on different levels, such as the system level, architectural level, the logic/RT level, the gate level, the circuit/transistor level and the physical/technology level, et al. Using low power design techniques at different levels can approach different power savings. Generally, the high level design methods have more effect on power saving.

Power or energy optimization can also be performed at different stages in the design process and may address different targets such as dynamic power or static power. In the following, we will briefly introduce the main low power design methodologies realized at design time, runtime and standby separately. We should always be aware of the fact that low power design usually calls for employing a combination of various low power techniques.

1. Reducing the Supply Voltage

As mentioned in formula (7-5), dynamic power dissipation depends quadratically on the supply voltage V_{dd}. Voltage scaling is therefore the most attractive and effective method for low power design. However, when supply voltage is lowered, it comes at a cost: the delay of the CMOS logic gate increases. The loss of performance can be compensated by some architecture and logic optimizations, such as parallelism and pipelining.

Using parallel processing structure means build $N(N \geqslant 2)$ same functional module instead of one, and then N problems can be solved concurrently. Therefore, each module can now operate at the $1/N$ speed with the system keeping the same throughput, and the clock frequency can be slowed down to $1/N$. The relaxed delay requirement enables a reduction of the supply voltage without depressing the system performance. Fig. 7.4 gives an example of parallel design.

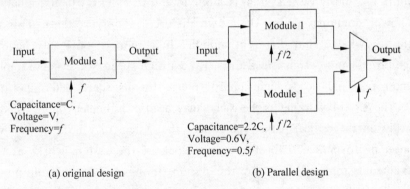

Fig. 7.4 Parallel structure

The power consumption of the original design and the parallel design can be computed by the following equations:

$$P_{original} = C \cdot V^2 \cdot f \tag{7-7}$$

$$P_{parallel} = 2.2C \cdot (0.6V)^2 \cdot (0.5f) = 0.4 P_{parallel} \tag{7-8}$$

Through the comparison of equation (7-7) and (7-8), we can see the effect of parallel structure in reducing power. The main shortcoming of this method is the substantial area overhead.

Pipelining is another method introducing concurrency. This technique divides the critical paths by inserting extra registers between the neighboring modules. Due to the reducing of delay, the system is now able to work at lower supply voltage with keeping the throughput. Fig. 7.5 gives an example of using pipeline in a design. Compared to the fact that parallel structure usually increases the area by more than 100%, which is often not acceptable, the pipeline structure can gain about the same power benefits with much smaller area overhead. But it also has some shortcomings, such as increasing latency and design complexity.

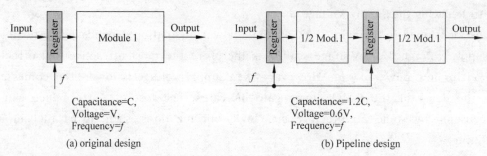

Fig. 7.5 an example of using pipeline

2. Using Multiple Supply Voltages

Lowering the supply voltage of selected blocks in the system can also help reduce power consumption significantly. This strategy using multiple supply voltages in a design is often called Multi-V_{DD}.

In a Multi-V_{DD} design, the system is subdivided into blocks, which are called voltage regions or power domains, having their own different supply voltages. This method is based on the knowledge that different modules in a modern SoC may have different performance targets and design constraints. Take a high speed SoC as an example, the DSP core runs most fast and needs high supply voltage for its speed determines the system performance. The SRAM, working as the cache, may need even higher supply voltage, because they are usually on the critical paths. The rest of the system may need much lower supply voltage to meet performance. Therefore, each block of the system works at the lowest supply voltage consistent with meeting the system timing, and remarkable power saving can now be achieved. Fig. 7.6 gives a framework of a Multi-V_{DD} SoC design.

However, using Multi-V_{DD} adds some complexity to the design. For example, we need a more complex power grid and level shifters (or level converters) on signals running through blocks with different V_{DD}.

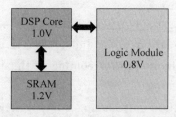

Fig. 7.6 Multi-voltage architecture of a SoC

3. Using Multiple Device Threshold

As explained earlier, when process geometries shrunk to sub-100 nm, leakage power became a big problem. Using libraries with multiple threshold voltages has become an efficient strategy bringing leakage reduction without any dynamic power or performance costs.

Many libraries today offer two or three versions of their cells: Low threshold, Standard threshold, and High threshold. The high threshold cell features a leakage that may be about one order of magnitude lower than that of the low threshold cell at the expense of some performance loss.

Introducing multi-threshold technique has little impact on the normal design flow. No level shifters or any other special circuits are required when using multiple thresholds. The designers only need to increase the thresholds in timing path which are not critical. Modern EDA tools can usually support multi-threshold design very well.

There are two strategies can be used for multi-threshold design. If minimizing the leakage power is more important than achieving a minimum performance, the design can start from high threshold library followed by swapping in low threshold cells. Otherwise the design process can go the other way around, which means synthesizing with the low threshold library first and then replacing the cells not on the critical path with their high threshold equivalents.

4. Dynamic Voltage and Frequency Scaling(DVFS)

Fixed supply voltage reducing often conflicts with performance requirement, which is unacceptable. As we know, a system or parts of the system not always work at the highest speed. When the workload decreases, it is possible to lower the clock frequency to reduce power. But by adopting only this method, which is named Dynamic Frequency Scaling(DFS), the total energy consumed for a task will not be saved because every switching occurs still at the high supply voltage. However, if we afford lower supply voltage when reducing the clock frequency, it will lead to both dramatic dynamic power reduction and energy saving.

To achieve substantial energy saving while maintaining the required throughput for peak workload, both the frequency and the supply voltage should be dynamically adjusted based on the current workload of the circuit. This technique is usually called Dynamic Voltage and Frequency Scaling(DVFS).

DVFS also brings some challenges to the system design. Because DVFS is closely relevant with the workload being executed, measuring and predicting the workload accurately, which is usually implemented by software, is extremely important before scaling the voltage and frequency. Any inaccurate estimation will greatly decrease the efficiency of DVFS. What is more, changing the supply voltages introduces some time delay and energy overhead, which can not be ignored. Last but not the least, an additional off-chip voltage regulator is needed for the processor or each voltage scaled parts of a

SOC, and a voltage level shifter is required whenever a low voltage signal is driving a high voltage receiver.

5. Dynamic Threshold Scaling(DTS)

Equation (7-6) shows that a linear change in threshold voltage will bring an exponential change in sub-threshold leakage current which dominates the static power in previous technology nodes. In order to reduce the leakage power, instead of adjusting the supply voltage and frequency in DVFS, the Dynamic Threshold Scaling(DTS) controls the threshold voltage according to the current workload, by controlling the body bias. This technique is especially attractive in light of the increasing impact of leakage power.

Every MOS transistor has a fourth terminal(the substrate), which can be used to scaling the threshold voltage by reverse or forward biasing. For high speed computation tasks, the threshold voltage should be adjusted to the minimal value; for the background tasks or high-latency tasks, which can be executed at a reduced frequency, the threshold can be changed to a higher value to reduce leakage; when the system is in the standby mode, the threshold should be set to the highest value.

It is alluring that the DTS technique does not change the circuit topology, does not need any voltage level shifter, and comes without any performance decrease. Generally speaking, DTS is easer to be implemented than DVFS. But the drawback is that it is required to control all the four terminals of both NMOS and PMOS transistors independently, which can only be realized in the triple-well technology. Though it is proved that DTS strategy is effective in reducing leakage power for 70nm process, its effectiveness is rapidly decreasing with the scaling of the technology below 100nm because the range of threshold scaling by dynamic body biasing is limited.

6. Operand Isolation

Operand isolation is often used to save dynamic power dissipation in data-path by reducing the unnecessary switching activity. Today's digital circuit designs usually contain many data-path modules that only occasionally perform useful computations but spend a remarkable amount of time in idle states. However, switching activity at their inputs in their idle states leads to redundant computations which are not used by the downstream circuits. The power dissipated by the redundant computation activity is just a great waste!

Operand isolation was first introduced in IBM PowerPC 4xx-based controllers. The main idea of this method is to identify the redundant operations and, utilize special isolation circuitry, prevent switching activity from propagating into a data-path module whenever it is about to perform a redundant operation. Therefore, the transition activity of the module is reduced significantly, resulting in lower power dissipation.

Operand isolation technique is very useful in applications which employ complicated combinational modules, such as ALU, long word adders and/or multipliers, and hierarchical combinational cells. DSP/CPU is a good example of the applications. In a particular clock cycle, usually only one of the data-path modules of the DSP/CPU executes the useful computation, other modules are just idle. Therefore, operand isolation can be used to control the redundant computation activity and reduce dynamic power significantly.

Operand isolation circuits can be easily inserted by using modern EDA tools. However, one important issue which should not be ignored is the leakage power. We should ensure that the isolation circuitry is designed and used in a way that the isolated module consumes minimal leakage power as well.

7. Clock Gating

In many systems, especially mobile application systems, standby (or sleep) mode takes great part of the operation time. Power management for this mode is critical for the battery life. In standby mode, total or parts of the system have no useful computational tasks to execute, and the dynamic power and standby power are hoped to be zero or very low. Many techniques are proposed to control the power dissipation in during period of inactivity.

In the digital circuits, the clock signal is usually highly loaded. The power consumed by clock network (often a clock tree) contributes a great part of the total dynamic power. Especially in standby mode, the clock switching is the main source of the dynamic energy consumption. In addition, when the registers receiving the clock maintain the same value, they will still dissipate some dynamic power.

The clock gating technique stops the clock signal fed into the idle modules, so it can significantly save the dynamic power not only from driving the clock tree but also from the unnecessary gate activity. Clock gating can be used in parts of a system or over the total circuit.

Nowadays, most of the libraries include specific clock gating cells. Meanwhile, modern EDA tools can automatically identify circuits where clock gating cells can be inserted without affecting the function of the logic. What is more, clock gating has many other advantages, such as low hardware overhead and little performance penalty. All of these make clock gating a simple, reliable and most widely used method of reducing power consumption.

8. Power Gating

After adopting clock-gating method, the power dissipation in standby mode is mainly from leakage, which may decide the battery life of portable devices and should not be ignored in today's sub-100nm technology. The power gating technique reduces both dynamic power and leakage power by cutting off the power supply of the idle blocks of a design, which is an important advantage over clock gating. For the systems or sub-systems in standby mode for long period of time, this technique is extremely attractive.

However, power gating is much more difficult to implement than clock gating and brings higher costs. One of the main challenges for power gating is the design of the proper power switching fabric. The common approach is to use the large size sleep transistors, either PMOS or NMOS devices of high threshold voltage for low leakage, to switch off the power supply rails when the circuit is in standby mode. The PMOS sleep transistor, often called "header switch", controls Vdd supply; and the NMOS sleep transistor, often named "footer switch", is used to switch GND supply. In real design, either header switch or footer switch, or both, can be used to implement the power switching fabric. However, in designs of sub-90 nm process, either PMOS or NMOS switch is only used because of the area penalty of the large sleep transistors and the constraint of the sub-1V power supply. Fig. 7.7 illustrates the switches in power gating design. The sleep transistors must be carefully optimized so that the benefits of leakage power saving overwhelms the area and power penalties from sleep transistors.

Compared to clock gating, another shortcoming of power gating is that it will take some time for a system or block to enter and leave the power gated leakage saving mode, which introduces additional delays and power penalty. What's more, the power gating method affects the inter-block communication of the system, and reduces the noise immunity if care is not exercised when designing the sleep transistors.

Other challenges in power gating designs include: design of isolation cells between sleep blocks and active blocks, design of the power gating controller, selection and use of retention registers for data and states, and so on.

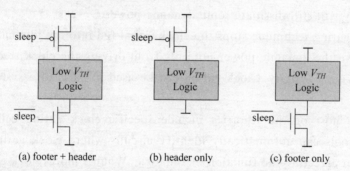

Fig. 7.7　Switches in power gating design

9. Dynamic Power Management (DPM)

Processors, such as DSPs and CPUs, are special digital circuits with good flexibility, and their operating mode may change greatly over time in a given application.

DPM, mainly used in processors and SoC, is a low power design methodology, which utilizes the least active components to meet the need for different computation tasks, by dynamically configure the system parameters. DMP encompasses a set of techniques to selectively shut off the components in idle mode, achieving a highly energy-efficient system.

DPM needs the support of hardware and software. Software predicts whether a device can go to sleep long enough to save energy, mainly based on two kinds of policies, which are predictive and stochastic. Hardware support mainly includes operating mode management and some basic low power method, such as clock gating, DVFS, power gating.

The basic idea of the operating mode management is to cutting off the clock signal or power supply to the circuits when they are in sleep mode. Besides active mode, the most common low power operating modes(usually called sleep modes) in processors are: idle mode, standby mode and power down mode. In sleep modes, the frequency or supply voltage can be reduced, and the modules that are not invoked are hung to save power dissipation.

7.3 Physical Level of Low-Power Design

In most of the applications, such as cell phone, audio-processing, video-processing, systems operate in burst mode, which is working in high performance mode for a short time following with a long interval in idle mode. The working mode is called active state of system, while the idle mode is called standby.

Furthermore, even the system is in active state, it doesn't mean that the system must work in the highest performance all the time. During design stage, the designers must make sure that the DSP can afford the maximum workload. However, the DSP usage is very low during most of the time. So, chance is coming. When workload is not the highest, there is some space for power optimization by doing some trade-off dynamically between performance and power.

According to the description above, the standby mode has gained a lots attention for its large overall power budget. Ideally, the dynamic power consumption during standby period should be reduced to zero through careful design. Furthermore, the static power should also be very small. However, the static power consumption is always hard to cut off with advanced technology scaling. The focus of this section is to discuss several methods to eliminate both dynamic and static power dissipation during standby period.

1. Clock Gating

All the switches of a subsystem in standby mode are the main source of dynamic power consumption, which is function meaningless, including clock tree and data propagation. So clock gating technology is implemented with the data isolation, as shown in Fig. 7.8.

The Structure of Clock Gating includes:

(1) Isolate the global clock tree from clock tree in idle subsystem. With clock tree isolated, no switching activities happen and the clock tree load capacity is eliminated efficiently. Thus, the power consumed by the clock tree can be removed.

(2) Isolate the input of standby subsystem from the output of the pre-subsystem. When the subsystem's clock tree is isolated, the activity of the inputs also brings on some additional power consumption in the combined logic. Isolation between pre-module

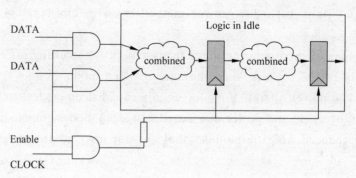

Fig. 7.8 Structure of the clock gating

output and input of this module can efficiently eliminate this part power.

That how and where to implement the clock gating has a vital impact on the submodule in standby mode.

(1) Shut down the clock tree of sub-module.
(2) Shut down the clock tree of the system.
(3) Shut down the clock tree of the system and the clock generator.

So, kinds of work mode exist because of different management methods for system clock, such as multi-mode supported in today's DSP Core:

(1) All the sub-modules of the system work.
(2) Part of the clock networks are shut down.
(3) The clock networks of all sub-modules but timer are shut down.
(4) Shut down all the clock network of system.

Clock gating is one of the most widely used low power methods. It is compatible with the standard design flow and EDA tools. In addition, the area penalty of clock gating is negligible. Clock gating technology can be implemented either in RTL level, written by designers, or in gate level, generated by design synthesis tools. Today's logic synthesizer can easily insert clock gating module into design wherever it finds a multiplexer logical structure. However, the tools can't implement global clock gating scheme. Designers can directly instantiate clock-gating logic cell into RTL code to implement global clock gating.

However, it has some impact on the system clock tree, increasing clock skew. Further more, the operand isolation may worsen the delay of the critical path. In addition, clock gating causes the load on the clock network varying dynamically, which may introduce another noise source into system.

The challenge of the clock gating is how and where to implement in the system. According to the usage granularity, there are two methods: grain granularity and coarse granularity. Coarse granularity is the most widely used now. Additionally, whether a sub-module is suitable for using clock gating, it depends on how long the standby mode lasts.

2. Power Gating

When clock gating is used for the sub-modules in standby mode, the switch power can be reduced nearly to zero. However, the static power remains. The power gating is cutting off the supply of the idle sub-module, reducing the static power to zero. It inserts switch transistor between the global power network and the virtual power network, which is the local power of the idle module. So, when the switch transistor is open, the sub-module works as usual. When the switch transistor is off, the submodule also is cut off. The structure of the power gating technology includes: the switch network, input and output isolation cell and state retention registers.

As shown in Fig.7.9, a normal structure of SoC include: DSP Core, Cache, BUS, peripheral devices, external memory and clock generator. An additional module in the Fig.7.9 is the power gating controller, which is especially designed for the power gating. The power network of the system is divided to three parts: VDDSOC, VDDCPU, VDDRAM. VDDSOC is always on, powering the parts which are always working, such as clock generator, power controller and peripheral devices. VDDCPU powers the DSP Core and the Cache. VDDRAM supplies power to the external memory.

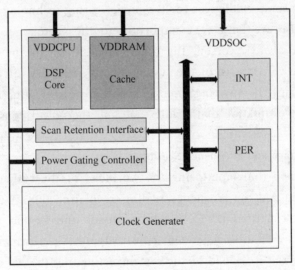

Fig. 7.9 The Structure of the Power Gating

The working state of a system, which has taken the power gating technology, can be divided to multi-level:

(1) Shut down the clock of DSP core, however keeping the supply on.

(2) Cut off the power supply of the DSP core, keeping the key states by the state retention register. Cache works as usual. The system can be awakened in the least time.

(3) Cut off the power supply of the DSP core, transferring the key states to the external memory through the scan register chains. Cache works as usual. The system can be awakened in the less time.

(4) Cut off the supply of the DSP core, Cache, transferring the key state to the external memory through the scan chains. When the system awakened, the content of the

cache must be rewritten by instructions. The awakening time is the longest.

Power gating is a key method which must be implemented in the ultra low power system. It is compatible with the standard design flow and EDA tools. However, it should be noted that power gating is not suitable for all application. For the application without a significant idle time, the power gating technology may not make any sense. The case may get worse for the power penalty by frequently turning off/on switch transistor. Designer should also notice that the implementation of the power gating makes the synthesis and the backend of the system more complex, especially for the power network and the clock verification.

Compared with the clock gating, the penalty of the power gating is more serious. Considering the area, the power gating needs many state retention registers and the isolation cells. Sometimes the area of the system using grain granularity can be 2~4 times as before. As for the performance, the introduction of the switch transistors reduces the supply voltage, worsens the critical path.

There still exist some challenges for the power gating technology. The awaking time and the voltage pulse are both very important to the switch network. However, the improvement of one always worsens the other. In addition, the power gating makes it more difficult to the verification of the system. Lastly, it is very difficult to estimate the power consumed during the awaking time. So, whether power gating is suitable for a sub-module is a result of estimation of the idle time and the power consumption during the awaking time.

3. Dynamic Voltage and Frequency Scaling

DVFS is a method that monitoring the workload of the system, then dynamically managing the voltage and the frequency scaling, regulating the system state, to achieve the aim of power reduction. DVFS needs to divide the system into several power domains. Each works with an independent power supply. A performance monitor module is added, monitoring the workload of the system, estimating thereal delay needed of the critical path, then transmitting the messages to the voltage management module and the clock generate module, regulating the voltage supply and the clock frequency of the sub-module.

The structure of the DVFS is shown in the Fig.7.10. The system is divided into several power domains: ①DSP Core, Cache, CPU performance monitor module, powered by VDDCPU; ②Memory, powered by VDDRAM; ③BUS, BUS performance monitor module, clock generate module, self-adjust voltage controller module, PLL, powered by VDDSoC.

Supposed the system works in a normal voltage and clock frequency. When the performance monitor module detects the fact that performance of the system can not meet the need of the workload, the voltage of the CPU should increase. Firstly, lock the frequency of the CPU to the current frequency. Then, increase the voltage of the CPU

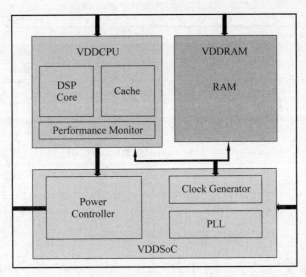

Fig. 7.10 The structure of the dynamically voltage and frequency scaling

until the performance need of the workload can be met. Lastly, transmit the frequency to the target frequency. Vice, versa.

DVFS technology is compatible with the design flow and EDA tools, widely used in the IC design. It makes good use of the performance needs of the system workload, achieving the trade-off between performance and power in real time, resulting in the power reduction. More important, the implementation of DVFS has little impact to the highest performance of the system.

However, When implementing DVFS in a system, performance monitor module, power manage module, clock generate module should be designed especially. These make the design of the system more complex. Additionally, the levelshifter between the different power domains may worsen the delay of the critical path.

There also exist some challenges for DVFS. How to estimate the workload of the system accuracy has a greatly impact on the efficiency of DVFS. How to generate suitable voltage and frequency to the system is another serious problem, which also has a greatly impact on the optimum system state.

7.4 Technical words and phrases

7.4.1 Terminology

Sub-threshold Leakage(ISUBTH)	亚阈值泄漏
Gate Leakage(IGATE)	栅泄漏
Gate Induced Drain Leakage(IGIDL)	栅致漏极泄漏
SRAM	静态随机存取存储器
Dynamic Voltage and Frequency Scaling(DVFS)	动态电压和频率调整

续表

Dynamic Threshold Scaling(DTS)	动态阈值调整
Clock Gating	门控时钟
Power Gating	门控电源

7.4.2 Note to the text

(1) Dynamic power consumption mainly consists of switching power and short circuit power, of which the former dominates. Switching power, known as capacitive power, is consumed during charging and discharging the parasitic load capacitor.

动态功耗主要包括开关功耗、短路功耗，其中开关功耗占统治地位。开关功耗作为容性功耗，主要是在寄生负载电容的充电和放电过程中消耗的。

(2) Because of the finite slope of the input signal during switching, the PMOS and NMOS devices are simultaneously turned ON for a very short period of time during a logic transition, allowing a short-circuit current to run from V_{DD} to ground.

由于输入信号在开关导通过程中具有有限的斜率，在一小段时间内 PMOS 和 NMOS 器件在逻辑过渡过程中会同时导通。这会产生从 V_{DD} 到地的短路电流。

(3) Even worse, sub-threshold leakage current increases exponentially with temperature, which makes the leakage power easy to exceed the design constraint at the worst case while it is acceptable at room temperature.

更为严重的是，亚阈值泄漏电流随着温度呈指数增加。这会使得泄漏功耗很容易超过室温下可接受的最坏情况时的设计约束。

(4) Power or energy optimization can also be performed at different stages in the design process and may address different targets such as dynamic power or static power.

我们可以在设计过程中不同阶段进行功耗和能量优化，同时优化的目标也可能是动态功耗或者静态功耗。

(5) The clock gating technique stops the clock signal fed into the idle modules, so it can significantly save the dynamic power not only from driving the clock tree but also from the unnecessary gate activity.

门控时钟技术可以防止时钟信号注入到闲置的模块中，因此它可以从时钟树驱动和不必要的逻辑门活动性中节约大量动态功耗。

(6) DPM, mainly used in processors and SoC, is a low power design methodology, which utilizes the least active components to meet the need for different computation tasks, by dynamically configure the system parameters.

动态功耗管理技术主要用于处理器和 SoC 中,它是一种低功耗的设计方法。这种方法利用最少的工作器件,通过动态配置系统参数来满足不同的计算任务需求。

(7) It makes good use of the performance needs of the system workload, achieving the trade-off between performance and power in real time, resulting in the power reduction.

动态电压和频率调整充分利用了系统工作负载的性能需求,实现了实时性与功耗之间的折中,降低了功耗。

参 考 文 献

N N Tan,D M. Li,Z H Wang,Ultra Low Power Integrated Circuit Design-Circuit,Systems. and Applications. Springer Press. NewYork,2014,67-78,97-101